# Sith, Slayers, Stargates, + Cyborgs

# Popular Culture & Everyday Life

Toby Miller
*General Editor*

Vol. 19

PETER LANG
New York • Washington, D.C./Baltimore • Bern
Frankfurt am Main • Berlin • Brussels • Vienna • Oxford

# Sith, Slayers, Stargates, + Cyborgs

*Modern Mythology in the New Millennium*

Edited by David Whitt + John Perlich

PETER LANG
New York • Washington, D.C./Baltimore • Bern
Frankfurt am Main • Berlin • Brussels • Vienna • Oxford

**Library of Congress Cataloging-in-Publication Data**

Sith, slayers, stargates, & cyborgs: modern mythology in the new millennium /
edited by David Whitt, John Perlich.
p. cm. — (Popular culture and everyday life; v. 19)
Includes bibliographical references and index.
1. Mythology. 2. Technology—History—21st century.
3. Popular culture—History—21st century. I. Whitt, David. II. Perlich, John.
III. Title. IV. Title: Sith, slayers, stargates, and cyborgs.
BL312.P47 791.43'615—dc22 2007026678
ISBN 978-1-4331-0095-6
ISSN 1529-2428

Bibliographic information published by **Die Deutsche Bibliothek**.
**Die Deutsche Bibliothek** lists this publication in the "Deutsche
Nationalbibliografie"; detailed bibliographic data is available
on the Internet at http://dnb.ddb.de/.

Cover design by Joshua Hanson


29 Broadway, 18th floor, New York, NY 10006
www.peterlang.com

# Table of Contents



## Mythology and Biology: Sith and Cyborg

## New Frontiers: Cylons and Stargate; *Serenity* and The City

## Critiquing the Myth and Looking to the Future: Slayers, Starfleet, and the Schwartz

# Acknowledgments

A project like this cannot come to fruition without help from numerous sources. The editors would like to thank friends, family, and loved ones for continuous support and interest in the project. Special thanks to Dr. Jessica Henry, Caiden Henry-Perlich, and Kelyn Henry-Perlich for giving "Daddy" the time to complete this book. Hastings College (specifically Dean Rich Lloyd and the members of the Faculty Affairs Committee) must also be thanked for support in the completion of this work.

David would like to extend special thanks to John Doan at Trade-A-Tape in Lincoln, Nebraska, for his assistance tracking down *Tales of the New Teen Titans #1*. Additional thanks to the Communication and Theatre Arts Department at Nebraska Wesleyan University and Byron Bonsall for his wealth of comic book knowledge and encouragement.

Both editors would like to extend deep appreciation to the co-contributors of this volume. When we started this journey we had no idea where the path would lead us. Our fellowship has been enjoyable and we gladly would travel this road again. On behalf of each author, we thank the families, friends, and academic allies who made this project possible.

Finally, we wish to thank Mary and Toby with Peter Lang for supporting this project; and Sophie Appel (also at Peter Lang) for her constant guidance and encouragement.

•CHAPTER ONE•

# Prologue: Not so Long Ago...

*John Perlich and David Whitt*

Science Fiction and Fantasy are our New Mytholcgy.

—S. Voytilla (1999, p. 260)

From the small seed grows a mighty tree; the book you are reading is an affirmation of this adage. Like Pegasus, born from Medusa's blood, this book has unusual roots. Our genesis was unintentional (if not accidental) and at no point did we hope to sow the seeds of Yggdrasil or excavate the tendrils of Olympus. Instead, we began in a moment similar to the one experienced by celebrated scholar of myth Joseph Campbell when he stood on the steps of a subway station emerging into midtown Manhatten and proclaimed, "The latest incarnation of Oedipus...stands on the corner of Forty-second Street and Fifth Avenue, waiting for the traffic light to change" (Campbell & Moyers, 1988, p. xiii). Like Campbell, each of our contributing authors perceives mythology as a contiguous force existing simultaneously within us and yet beyond the constitutive world. Our observation of mythological moments comprises the substance of this book.

As mundane as it might seem, the origins of this project can be traced to a late afternoon National Communication Association (NCA) forum in Private Dining Room One on the third floor of the Hilton Chicago in 2004. The room normally seats seventy, but it was standing room only for a NCA Town Hall Debate by the editors of this project (and two other participants) titled "*The Matrix, The Lord of the Rings,* and *Harry Potter:* Modern Mythology and the New Millennium." When the debate concluded, a group of audience members lined up to chat with us about their own research into mythology.

After the last spectator had finally left the room, we packed our notes in excited disbelief and wondered, "What just happened?"

Wrongly, we had assumed that our academic contemporaries might not understand the significance of our discussion. Unlike Bill Moyers' colleague who asked, "Why do you need mythology...all these Greek gods and stuff?"(Campbell & Moyers, 1988, p. xiv), our peers demonstrated an acute awareness of the energy in all that "stuff." On that cool November day, on the shore of Lake Michigan, we learned that mythology taps into something that indeed will appeal to a modern audience. As Campbell said, "So much of life today...from religion and war to love and death" is ritualized and mythologized (Campbell & Moyers, 1988, p. xiv). After the presentation we found ourselves wondering, "Why not develop the Town Hall Debate into an academic forum by encouraging others to submit their research as it applied to various science fiction or fantasy texts?" The result is the book you are now reading.

## Why Study Myth?

For many high school or college students, their introduction to classical mythology came from reading Homer's *The Odyssey* or *The Iliad.* For some, these stories were exciting epics, filled with heroic quests, fantastic creatures, and fierce battles. For others, these centuries-old narratives were like reading Shakespeare for the first time, with strange language, a myriad of characters, and lengthy exposition. Even if one has not read these literary classics, most people are familiar with the names Zeus, Athena, and Achilles or have heard of Sirens, Cyclops, or the Trojan Horse. Thousands of years after they were written, Homer's hymns still manage to inspire, having become perhaps the most celebrated mythic poems in history. However, Homer is but a prologue to the remarkable power and relevance of myth throughout civilization.

From the ancient temples of Rome, to the dense rainforests of South America, to the Native American plains, mythic stories of gods, goddesses, heroes, and creatures are a vital and vibrant part of every culture. In *The Hero with a Thousand Faces* (2004) Campbell writes, "Throughout the inhabited world, in all times and under every circumstance, the myths of man have flourished; and they have been the living inspiration of whatever else may have appeared out of the activities of the human body and mind" (p. 3). Campbell defines myth as "stories of our search through the ages for truth, for meaning, for significance...Myths are clues to the spiritual potentialities of human life" (1988, p. 5). Campbell's mythic canvas is incredibly broad, examining the gods and heroes of not only ancient Europe but also of China, India, and the Far East. He argues that through the analysis of myth, individuals can better understand their humanity and the experience of being alive.

While Campbell's influential body of work provides an essential foundation to understand the history and purpose of myth, other scholars have examined the value of these cultural narratives. Stock (1993) explains, "Mythology speaks to our deep longing to understand the human condition: to know where we came from, why we are here, what will become of us" (p. 240). Rowland (1990) states, "Myths are stories which symbolically solve the problem facing the society, provide justification for a social structure, or deal with a psychological crisis" (p. 103). He adds, "in terms of function, myths are the most important stories in society. Therefore, if a story is in fact mythic there should be strong evidence that it is taken **very** [author's boldface] seriously by the people who tell it" (p. 105).

The acceptance and appeal of myth make these stories not only culturally significant but also rhetorically meaningful. Rushing (1989) explains, "the important myths of society are complex blends of archetypal and rhetorical elements" (p. 2). Similarly, Hart (1990) contends, "Myth and rhetoric have a symbiotic relationship. Myth gives rhetoric something to say and rhetoric gives myth impact in everyday life" (p. 321). For example, the Frontier myth is an enduring vision of American individualism and exploration, applied to continental expansion and space exploration. In this instance, the Frontier myth serves as a binding cultural influence, its rhetorical power leaving an indelible mark on the American psyche. In this way myths are more than just simple or entertaining stories; they are, to quote Campbell (1988), "the homeland of the muses, the inspirers of art, the inspirers of poetry" p. 55).

While many myths provide a unique window to the past, are there more contemporary myths that provide insight into the future? On the eve of the twenty-first century Voytilla (1999) argued:

> Science Fiction and Fantasy are our New Mythology, and provide an important canvas that allows us to explore society's issues. The Journeys may propose the dangers we could face if an "evil" continued—e.g., nuclear power,

> genetic engineering, and technology. These Journeys may represent our society's phobias in a given era (the Cold War, nuclear proliferation, or prejudice); but they may also present our hopes of a utopian world in which we have overcome a contemporary issue, or simply celebrate our humanity. (p. 260)

Wagner and Lundeen (1998) explain that science fiction is mythic "insofar as it embraces the 'deep story' attributes of myth. Both science fiction and traditional myth are 'thought experiments' that posit a 'What if?' question and then pursue its consequences...both traditional myth and science fiction provide narratives that probe the unrealized possibilities of being" (p. 7).

Patai (1972) describes modern science fiction myths as eschatological, based on the branch of theology concerned with the destiny of humanity. He explains that science fiction mythology includes elements of struggle, battle, and victory that serve to reassure humanity about our future by stating, "The greatest victory that science fiction can envisage is that of man over the technological environment of his own creation" (p. 87). Rushing (1990) states these myths "symbolize a new stage in the evolution of cultural consciousness, in which scientific discovery is recontextualized as part of a greater whole" (p. 144).

In short, answers to life's great questions may, in fact, lie waiting for us in mythology (be it old or new). Invoking the memory of Campbell, the journalist Bill Moyers helps us understand the significance of all this "stuff" when he clarifies the role of mythology in the life of one brilliant scholar (Campbell & Moyers, 1988):

> To [Campbell] mythology was "the song of the universe," "the music of the spheres"—music we dance to even when we cannot name the tune. We are hearing its refrains "whether we listen with aloof amusement to the mumbo jumbo of some witch doctor of the Congo, or read with cultivated rapture translations from the sonnets of Lao-tsu, or now and again crack the hard nutshell of an argument of Aquinas, or catch suddenly the shining meaning of a bizarre Eskimoan fairy tale." (p. xvi)

The diversity of subjects, themes, and insights found in this book mirrors Campbell's generalist philosophy that makes connections among a variety of disciplines, sources, and philosophies. For readers who find they are still wondering, "Why should we care about myths?," we also invoke Campbell and retort, "Go on, live your life, it's a good life—you don't need mythology...but...you may find with a proper introduction, mythology will catch you" (Campbell & Moyers, 1988, p. 3).

In the chapters that follow you will be exposed to a wide array of discussions of what we describe as New Mythology in contemporary science fiction fantasy texts. The first section of the book—"Mythology and Biology"—explores issues in genetics, cybernetics, "the human spirit," and mythology. The second section—"New Frontiers"—explicates several extensions of mythic frontier themes. The final section of the book—"Critiquing the Myth and Looking to the Future"—offers critical perspectives on new paths in mythology. Most of our authors deconstruct an artifact and illustrate the powerful magic that can be conjured when one listens to the "music of the spheres"; some authors also demonstrate the folly of those who are seemingly deaf to the "song of the universe." The book is not without controversy, and in keeping with the spirit of the critical tradition, one of our goals is to facilitate dialogue regarding the universal appeal of myth and mythology. We ultimately hope that this first chapter has provided "the proper introduction" and planted a seed of interest. Keep reading and you may find, as Campbell suggests, "mythology will catch you."

## References

Campbell, J. (2004). *The hero with a thousand faces.* Princeton, NJ: Princeton University Press.

Campbell, J. & Moyers, B. (1988). *The power of myth.* New York: Doubleday.

Hart, R.P. (1990). *Modern rhetorical criticism.* New York: Harper/Collins.

Patai, R. (1972). *Myth and modern man.* Englewood Cliffs, NJ: Prentice-Hall.

Rowland, R. C. (1990). On mythic criticism. *Communication Studies, 41*(2), 101–116.

Rushing, J.H. (1989). Evolution of "The new frontier" in *Alien* and *Aliens*: Patriarchal co-optation of the feminine archetype. *Quarterly Journal of Speech, 75*, 1–24.

Rushing, J.H. (1990). On saving mythic criticism—a reply to Rowland. *Communication Studies, 41*(2), 136–149.

Stock, G. (1993). *Metaman: The merging of humans and machines into a global superorganism.* New York: Simon and Schuster.

Voytilla, S. (1999). *Myth and the movies: Discovering the mythic structure of 50 unforgettable films.* Studio City, CA: Michael Wiese Productions.

Wagner, J. & Lundeen, J. (1998). *Deep space and sacred time:* Star Trek *in the American mythos.* Westport, CT: Praeger.

# Mythology and Biology:
# Sith and Cyborg

# CHAPTER TWO

# "I've Got a Bad Feeling About This…": Lucas Gets Lost on the Path of Mythos

*John Perlich*[*]

In the fall of 2004 several friends and colleagues joined me at the annual proceedings of the National Communication Association. We were there to catch up on old friendships, share tales of academia, and ultimately present a panel on new mythologies and the role that the science fiction/fantasy genre might play in the co-constitution and co-construction of myth and mythic themes. Each of the presenters on this panel was surprised at the turnout for what we considered a pop-culture topic; admittedly, we assumed that our fellow scholars would consider the artifacts we were exploring to be passing fads (at best). All of us noted that this panel was easily the best-attended session we had ever participated in (impressive because, together, we had probably been involved in at least one hundred sessions). Perhaps we had indeed struck a subconscious resonating cord that called out for attendance, some inherited and deeply etched response that provoked our audience to join us in the exploration of myth.

The format of the panel clearly begged for patronage from scholars with a sense of humor. Each of the presenters would be making a case for the "Best Exemplar of New Mythology" in the next millennium. Essentially, we had created an Oscars-type award panel for science

---

[*] Dr. John Perlich (M.A., Minnesota State University-Mankato; Ph.D., University of Nebraska) is an Associate Professor of Communication Studies at Hastings College in Nebraska.

fiction and fantasy fans. The red-carpet nominees included popular blockbusters (an intentional choice on our part) like *Harry Potter, The Lord of the Rings,* and *The Matrix.* Each panelist would make a case for their icon(s) and plead with the audience for sympathy and ultimately agreement.

It was my job to make an argument for Tolkien's mythology (recently revisited in the cinematic efforts of Peter Jackson) as the exemplar of New Mythology. My argumentation was heavily grounded in a type of rhetorical deconstruction known as generative criticism (Foss, 2004), a process through which the critic develops a rubric or schemata for analysis (because a fitting model in scholarship is neither readily available nor completely appropriate) and applies this framework to the artifact under scrutiny. Not surprisingly, I relied heavily on the work of Joseph Campbell (1904–1987) to develop a typology that would stand to characterize the essential elements of a mythic text.

Notably absent from our discussion was any sign of the mythical colossus created by George Lucas (known by most fanboys and fangirls as the "*Star Wars* universe"). The stunning silence of this multiverse seemed striking as I sat in my chair listening to my colleagues and waiting for my turn at the podium. I thought to myself, "How can we have a panel on New Mythology and not, somehow, include the mighty works of George Lucas?" Admittedly, this seemed to undermine the credibility and ethos of our presentation. The *Star Wars* universe had an undeniable impact on our culture. More than thirty years after the premiere of this pop-culture phenomenon, its impact can still be seen in retro T-shirts, videogames, role-playing games, fiction, television, etc. Even Joseph Campbell implied that George Lucas is the most important singular figure in New Mythology. Why was there no sign of Lucas on our panel?

I was certain that our audience would comment on this omission at the end of our presentations and decided (rather impromptu) that someone should address this apparent hole. So, during one of my colleagues' presentations (I hope he will forgive me), I found myself silently applying the constructs identified by Joseph Campbell in order to figure out how the *Star Wars* universe might measure up. To my surprise, I was quickly able to identify at least one serious flaw in the Lucas mythology. This "Achilles heel" could be used (for the sake of argument) to explain why several other modern mythologies might be more appropriate exemplars of neo-myth.

When it was my turn to speak, I began with a preface about the absence of Lucas' mythology from our panel. With Campbell as my sling and Tolkien as my stone, I struck a blow at "the Beard's"[1] work and was surprised that my attack seemed to stun this Goliath of science fiction. I knew that my blow had impacted soundly because it received murmurs of accord and nods of approval from my audience; clearly I was not the only person in the room with a critical attitude toward the mighty force created a long time ago.

Though my critique was ad hoc and relatively brief, I promised the audience that I would dedicate a future project to an in-depth articulation of my position—consequently, this chapter is a vow that comes to fruition. Its purpose is to make room for "other" new mythologies (as *Star Wars* is often seen as the heir apparent) by, first, confessing my relationship as more than a casual consumer of the *Star Wars* mythos, then explaining where Lucas seemed to "get it right" with regard to mythic appeals, and finally exposing the soft underbelly of this pop-culture giant.

## A Long Time Ago...

My personal relationship with *Star Wars* began when I was eleven years old in 1977. My mother wanted to take us to a movie and she mentioned a "space-western" that was getting lots of attention. I was not a fan of cowboys. With regard to outer-space, my interests were more terrestrial (dinosaurs, actually). So I did not jump at the offer to see the movie. However, I was outvoted by Mom and my brother, so I bit my tongue and went along for the ride. Hours later, when we walked out of the theater, my jaw was still dragging on the ground.

It is important for the reader of this chapter to know that I am not thrilled about Lucas-bashing. In fact, I loved the *Star Wars* universe growing up and only recently have become disillusioned with the double trilogy of films that make up the core material for fans. Following my first experience with *Star Wars: Episode IV: A New Hope* (*ANH*) in 1977, I became a collector and consumer of all things *Star Wars*. Something in the first film triggered a response in me, and from that moment I became a fan and collector. So many of the themes in the first trilogy (subjects discussed later in this chapter) resonated with me in a way that I could not begin to describe and explain. I watched *ANH* twenty-five times before the second film—*The Empire Strikes Back* (*ESB*)—was released on May 25, 1980.

As a collector of *Star Wars* merchandise, I had few peers in my age group. I saved up my allowance to buy each and every major action figure and play-set that came available. *Star Wars* became my religion, and Kenner became my church (at least it provided the idols of my polytheism). There was simplicity and purity in the themes of *Star Wars*, and it was almost comforting that these seemingly self-evident truths could be re-created in my home with the help of 3 ¾-inch replicas of my favorite protagonists and antagonists.

Almost as a sign of my faith, I continued to collect *Star Wars* merchandise well beyond the years that one would play with toys. Scholars such as Shelton (2006) would aptly call my attention given to these objects something akin to the fetishism of commodities, which has been well scrutinized from a Marxist perspective. The mystical power known as "the Force"—an omnipresent, transcendental spirituality that bound all creatures and objects together—seemed simple yet poignant and prolific. Most powerful for me was the theme of redemption, that those who walked a dismal path might be saved and essentially purified. The fact that the *Star Wars* universe offered rectitude and solvency for troubles that occurred in my personal life strengthened the connection to this phenomenon.

By the time I left for college, my *Star Wars* collection included nearly every action figure issued by Kenner, including some of the rarest. Through my college years and into graduate school my collection was stored (by gracious parents) or, after a mishap with my father's garage roof, kept with me through several moves. Even though they were no longer displayed, the tokens of the past remained important pieces of my life and ultimately traveled with me on my journey.

Almost twenty-five years after my first encounter with the *Star Wars* galaxy, I continued to assign meaning to the artifacts I had collected. In fact, as my wife and I prepared to buy a new house, my one request was a "man room," or "fortress of solitude" as one friend dubbed it. This room would prominently feature my *Star Wars* collection and allow me to display the objects that had been in storage for so long. However, even as I made plans to showcase the objects from my childhood, a nagging irritation was working at the back of my brain. The seed of discontent had been sown in May 1999 and was sprouting into a weed that would require attention.

Like many *Star Wars* fans, I waited for sixteen years while George Lucas pondered how he would set up his first three films in the trilogy, including *ANH*, *ESB*, and *Return of the Jedi* (*ROTJ*). I purchased my advance copy of what promised to be the birth of a new stage in the *Star Wars* universe, Terry Brooks' (1999) adaptation of the screenplay for *Star Wars: Episode I: The Phantom Menace* (*TPM*). Friends will tell you that I read the book in one sitting (a very late night), as promised. I had long been a fan of Terry Brooks (most famous for the *Shannara* series) and believed he would write a terrific adaptation.

After completing the novel, I was left with many questions but believed they would be answered with future installments to the *Star Wars* universe. Although I was confused by several developments in the *Star Wars* mythology, I hoped that these strange new phenomena would eventually be recast in the mold of the original mythos. Unfortunately, many of the new contributions to the *Star Wars* galaxy would not be contextualized, and consequently the seeds of discontent would be allowed to flourish.

Five years after the cinematic premiere of *TPM*, my readings of Joseph Campbell and a life of interest in pantheons from all ages of humanity provided me with the foundation to explain why the *Star Wars* galaxy seemed to be imploding. Although I agreed with Campbell's assertion that George Lucas had established himself as one of the most important contributors to New Mythology (specifically with his first trilogy of *ANH*, *ESB*, and *ROTJ*), I began to wonder whether a posthumous conversation with the author of *The Masks of God* might temper such high praise considering more recent contributions to the mythology. In fact, might Campbell's praise be curbed in the wake of *Star Wars: Episode I* (*TPM*), *II* (*Attack of the Clones* [*AOTC*]), and *III* (*Revenge of the Sith* [*ROTS*])?

The point of this chapter is not simple "Jar-Jar"-bashing. Nor is it my goal to pay homage to Kevin Smith's "Death Star" debate between Randal and Dante in *Clerks* (a pop-culture critique of the *Star Wars* universe). Instead, my hope is to scratch an itch by attending to a nagging concern regarding an epistemic shift in the *Star Wars* mythology. I believe this turn of events has significant implications for the longevity of Lucas' creative myth and ultimately for modern mythology.

I have disclosed my fleeting love affair with the *Star Wars* universe because it is an important methodological step in this project.

My goal in sharing the journey from fanboy to critic is grounded in phenomenological inquiry and is a process known as the epoché.[2] Because I reveal my biases and preconceived notions regarding the phenomena under investigation, readers can judge for themselves whether the essence of my argument is idiosyncratic and personal or actually meritorious on the basis of relatively independent facts. Thus, the reader of this chapter now knows (and will understand with further reading) that my initial biases toward the *Star Wars* universe were quite positive and that only recently have I become more critical of the mythic contribution of this artifact.

## Powerful Mythology

Joseph Campbell believed that mythologies exist, resonate, and endure because they tap into something similar to an innate releasing mechanism (IRM). An IRM can best be described as a theoretical biological process within us that responds to a stimulus in order to produce a preprogrammed action or pattern of behavior. There are many scholars of human behavior who believe that such preprogrammed patterns of action are either conditioned or inherited and exist within each of us. Within such bodies of research, the IRM is often renamed; other terminology for the IRM includes schemata or scripts.

Konrad Lorenz is among the most popular theorists of the IRM. To illustrate his theoretical premise (that some behaviors are triggered by deeply written programs), Lorenz (1937) analyzed the nesting behavior of the goose. According to Lorenz, when an egg rolls out of the nest, a gray goose will attempt to roll the egg back into the nest by arching its head over the egg and using the underside of its bill to retrieve it. Lorenz notes that a sweeping side-to-side motion of the head is used to balance the egg. Oddly, the gray goose continues the movement even when the egg has rolled away during the process—suggesting that something deep inside the brain of this animal causes it to continue this motion.

Similar to the deeply ingrained behavior of the goose is a pattern of behavior known as the "chicken dance." It is quite easy to "teach" a chicken how to move around because scratching for food is simply something the bird "does." Try, on the other hand, to train a chicken to stand still while it waits for food! Try your best to make a chicken stand still (response) for food (stimulus); it just won't do it. Some have argued that such patterns of behavior are learned (rather than biological or innate) and that "hard-wired" lessons are crucial for early survival. If the goose, for example, does not learn to follow its mother very early in life, it will not survive. Hence there have been quite interesting experiments on the imprinting process in such species. With regard to the human condition, some linguists have suggested that we must learn language very early in life and that if we do not it is difficult to acquire linguistic ability.

If language must be acquired early in order for us to communicate effectively with others, might the content of the message be as significant as the mode of communication? In other words, does the function of myth, legend, and story serve as a stimulus that will elicit something akin to the IRM? Joseph Campbell believed that mythology operates on a deeply ingrained cultural level and that myth functions as a stimulus capable of producing reactions within the conditioned person.

With regard to the study of communication and rhetoric, some theorists seem to agree with Campbell's argument; most notable among these scholars is Walt Fisher. Fisher (1987) articulated (via his Narrative Paradigm) that human beings are not always rational decision makers. While the classical school might suggest that human beings weigh the evidence and make decisions as a result of carefully considered deductive and inductive process, Fisher believes that we do not always make decisions as a result of linear logic. In fact, Fisher (1987) points out that many decisions are made according to how the consequence might fit within and/or fulfill a story or narrative (known as narrative coherence) or intuitive judgments (known as narrative fidelity). Fisher's paradigm helps us understand why people often make poor decisions (choices that fly in the face of logic). People often make choices because they anticipate their actions might fulfill a narrative script or story. For example, why would a chronic gambler spend money on a slot machine instead of clothes for the family or other necessities? There are many explanations for the behavior, but one possible conclusion is based on the function of narrative; the gambler believes that the lucky day has arrived (is predestined perhaps) and that winning the big payout will make all other (pleasant) dreams come true and fulfill the happy ending that we all often seek in life. In other words, the action of gambling is justified within the context of a narrative that concludes with the "happy ending."

Theories that directly or indirectly allude to instinctive or programmed behavior are not without critics. Lehrman (1953) points out several flaws within the more detailed postulates of the theory. However, Lehrman is never able completely to dismiss the fact that we often act in illogical ways for seemingly inexplicable reasons. Lehrman and other critics of IRM theory question the mechanics of the postulates but never completely succeed in rejecting the more fundamental premise of Konrad's position.

## The Mythology of *Star Wars*

So what role does Lucas play in this discussion of IRMs and narrative paradigms? As previously suggested, Joseph Campbell believed that mythology provides us with a guide and reinforces our learned patterns of expected behavior. Historically, it is likely that story and storytelling functions preceded actual language and that our ancient ancestors used nonverbal communication to share signification and, consequently, navigate a very cruel and inhospitable world. Though today our needs are no longer as primal, for obvious reasons stories, parables, legends, myths, and tales continue to play a role in shaping and guiding our lives.

From Campbell's perspective it seems that the basic role of mythology is to evoke and reinforce responses that have served us individually and collectively through time. For example, many tales throughout the history of humanity illustrate the perceived importance of the lesson from *Luke* 6:31, "and as you wish that others would do to you, do so to them." This quotation discourages Machiavellian acts of self-fulfillment and instead promotes altruism, generosity, and selflessness. Certainly this credo has helped the human species survive and thrive in difficult and trying times. Imagine a world in which the inverse of *Luke*'s quotation might hold sway: "Do so to them before they can do so to you." Most certainly the ranks of humanity would be thinner (at best) if not extinct (a.k.a. Social Darwinism).

The lesson from *Luke* 6:31 is illustrated in the Lucas mythology at the end of *ROTJ* when Luke Skywalker has an opportunity to kill his father, taking his place at the side of an emperor who has promised unimaginable power, glory, and position. Though Luke Skywalker is certainly justified in ending the life of his father for a long list of misdeeds (remember that this is a guy who killed Jedi toddlers and helped blow up an entire planet), Luke offers clemency to his father. In doing so, the mythic character Luke Skywalker reinforces a lesson passed down to us by other mythologies.

While there are countless themes and subthemes in the Lucas mythology that both reinforce and illustrate lessons handed to us through the ages, Campbell (1991) identified four essential themes as crucial to a true mythology; in concise form, these themes include appeals based on collectivity, individuality, cosmology, and the mystical/metaphysical. I have used the work of several philosophers, scholars, and theorists to frame these essential themes within the dialectical tradition (Altman, Vinsel, & Brown, 1981; Bakhtin, 1990, 1994; Baxter, 1994a, 1994b; Leinenweber, 1977; Maines, 1989; Montgomery, 1993; Rawlins, 1989, 1992; Rychlak, 1976; Werner & Baxter, 1994). By using the notion of "unified opposites" to make meaning of Campbell's (1991) themes, I have generated two dialectical thematic categories useful for analysis: collectivity/individuality and cosmology/metaphysicality.

## Collectivity and Individuality

Creative mythologies will stand the test of time if they appeal to both collective and individualistic themes. Dubbed by Campbell (1991) the social perspective, creative mythologies are likely to receive positive regard from their readers when they provide homage to social structures. As an example, in the creative mythologies of J. R. R. Tolkien, the Fellowship created in Rivendell and charged with the destruction of Sauron's ring provides a stage for unyielding friendships and the importance of unity and solidarity. In a truly mythic text, a protagonist who attempts to go it alone will most likely fail; hence the hero or heroine must seek the aid of others.

The first *Star Wars* film to demonstrate an appeal to collective perspectives was *ANH*. From the beginning, Luke Skywalker's journey and success were enabled by a wise mentor (Ben), strong princess (Leia), scoundrel (Han), benevolent monster (Chewbacca), and dutiful jester-servants (R2-D2 and C-3PO). Each of these characters played a crucial role in the path of the protagonist. It should be noted that the classic monomyth (explained in detail throughout this volume) places a heavy emphasis on collective structures at the beginning of a heroic adventure. Toward the end of a monomyth, we see a shift toward the second part of the dialectical construct—the individualistic perspective.

The individualistic perspective was recognized by Campbell (1991) as the psychological aspect—the centering and harmonizing of the individual. In traditional mythic structures this often involved the giving of oneself, and even the giving up of oneself altogether, to a greater cause. This stage in the monomyth could easily be characterized by Luke Skywalker's brief foray into the dark tree on the planet of Dagobah during his Jedi training. Luke is told by his instructor that he must enter the cave under the tree alone, bringing only what he takes with him and that weapons will not be needed. At this turning point within the *Star Wars* mythology, Luke Skywalker has been forged into a protagonist by relationships with friends, family, and mentors; now he must leave all of these structures behind and turn his gaze inward. The final test for a protagonist lies within the self. Luke soon learns on failing the test in the tree that evil resides within each individual; one wrong step and each of us is capable of walking the path of darkness. This lesson is made palpable when at the end of the test Luke succeeds in severing the head of Darth Vader; as the dark lord's faceplate falls from the helmet, Luke sees his own visage.

It is often the case that a protagonist loses himself or herself as a result of the journey in solitude toward the end of the monomyth. Frodo, for example, in J. R. R. Tolkien's creative mythology discovers that he cannot resist the power of Sauron's ring. The further he leaves the collective (losing touch with first Gandalf, then Aragorn, and finally Samwise), the closer he comes to the sacrifice of self. When he is momentarily alone, atop Mount Doom, he is at the highest point of tension from the psychological aspect—it is there, on this pinnacle, that he endures the ultimate test.

Within the *Star Wars* mythology, Anakin Skywalker aptly demonstrates that the further he travels in solitude on his journey toward heroic status, the deeper his loss of self. He first loses his mother, then shuns the teaching of his mentor, and ultimately questions the love of his princess. Lying alone, next to a lake of fire, Anakin learns that he is not powerful enough to save the galaxy by himself (a failure of hubris). Unlike Luke Skywalker, Anakin loses the battle for the self and must later be redeemed from darkness by his then-unborn son.

Campbell (1991) contends that powerful mythic texts resonate with the unified yet oppositional centripetal and centrifugal tensions between collectivity and individuality. Most scholars of human behavior would agree that human beings experience a compelling desire to belong within a collective yet achieve an identity that is independent of the collective or an identity that is achieved beyond the collective. The *Star Wars* universe succeeds in articulating this dialectical struggle and consequently exemplifies mythic text and argument within this thematic.

## Cosmology and Metaphysicality

Lucas successfully created what so few science fiction and fantasy authors are able to create—an impressively immersive material cosmos. Success in creating a mundane secular world that allows the reader/viewer to become absorbed in both story and character may prove to be the greatest achievement of George Lucas. Campbell suggests that those who resonate with mythic archetypes are able to present an image of the universe that is believable and terrestrial. Tolkien, for example, has often been charged with placing Mordor in a Eurocentric map (with Mordor located in approximately the same place as Germany). While Tolkien denied an intentional placement of Mordor in the center of a Euro-type cosmos, many readers identified with the realm of Middle Earth because of the cosmology created by the synthesis of author and reader.[3] Additionally, it is hard to deny that the genesis of Tolkien's mythos might have occurred during the days immediately after World War I and before World War II. By Tolkien's own admission he relied heavily on Anglo-Saxon and Gaelic mythos (among others) in his construction of Middle Earth; therefore it is not irrational to impose a Eurocentric frame when perceiving the Tolkien cosmos.

George Lucas created a cosmology almost perfectly suited to the template that Campbell called the monomyth. Literally isolated on a desolate desert planet, Luke Skywalker is begging for adventure. His righteous desire (a.k.a. superego) to serve others calls out subconsciously and pleads for him to leave his simple farm life. A spiritual mentor teaches him the way of mystical power and explains that the cosmos is both vast and compelling. On a journey toward actualization, the protagonist confronts dire opposition and finds himself entrenched in a battle between worlds in the heart of darkness.

Like the shadowy netherworld of the sidhe (Gaelic for elf or elven), the "Force" suggests and resonates with an implied mythic structure. The Greeks called this netherworld Hades; the Norse referred to their netherworld(s) as Asgard, Jotunheim, Alfheim; and King Arthur's wounds were tended in a land called Avalon. The parallel world implied by the Force in *Star Wars* lore suggests a realm that offers lucidity, longevity, and virility. Clearly there are two "worlds" in the *Star Wars* universe—the material world (or Core Universe) occupied by Tattoine, Coruscant, Dantoine, Yavin, and so on, and the spiritual world (or Transcendental Universe) ruled by the Force. It is implied on several occasions that while the Force can transcend its border, the Core Universe cannot stray from its material boundary.[4] Imagine a plane known as the Transcendental connected almost entirely by a grayish matter (like water or a fog)—the Force "surrounds us"; it "binds us." Imagine that creatures cannot travel from the Core Universe to the Transcendental Plane of the Force without supernatural power (or death), yet beings from the Transcendental Plane are able to travel back to the Core Universe under special circumstances. This one-way-door effect seems to be the dominant mode of operation in the *Star Wars* mythology.

In order to fulfill the mythic role, Lucas must succeed in creating a clear cosmology while distinguishing this material world from a transcendental spiritual plane. Admittedly, the cosmological universe created by Lucas and subsequent authors is impressive. In fact, like a map of Middle Earth, the *Star Wars* universe can be overwhelming in its intricacy. Unfortunately, the dialectical opposite of the cosmological (the transcendental) is not as fully developed in the *Star Wars* universe and provides the previously alluded to Achilles heel. Critics of my argument will point out references to the "Life Force" and "Unifying Force"; these are two examples of the fact that the Force exists on "more than one plane" (Brooks, 1999, p. 137). It is not the naming of these planes but rather the interaction of these planes with a material world that merits scrutiny.

When speaking about "planes of consciousness," Joseph Campbell (in Campbell & Moyers, 1988) might just as easily have been describing the Force from *Star Wars*. "There is the plane of consciousness," hypothesized Campbell, "where you can identify yourself with that which transcends...[it is] unnamable":

> It is transcendent of all names. The transcendent is unknowable and unknown. The mystery of life is beyond all human conception. We always think in terms of opposites—I and you, this and that, true and untrue—every one of them has an opposite. But mythology suggests that behind that duality there is a singularity over which this plays like a shadow game. (pp. 48–49)

Campbell's words resonate with the writings of the existentialist philosopher Martin Buber (1970), who describes the transcendental *I—thou* moment as beyond quantification and operationalization. Any attempt to explain the transcendent moment, Buber argues, spoils it and returns us to a state of reductionism. Campbell (in Campbell & Moyers, 1988) describes the tenuous nature of transcendence when he states:

> Every religion is true one way or another. It is true when understood metaphorically.[5] But when it gets stuck to its own metaphors, interpreting them as facts, then you are in trouble. (p. 56)

It is my argument that the most recent trilogy created by Lucas has, in fact, gotten him into trouble from a mystical perspective.

Although Lucas did succeed in evoking notions of parallel worlds such as Avalon, Hades, Heaven, Heliopolis, Olympus, Yggdrasil, and Asgard (and even Dante's Inferno), his immature treatment of transcendental and mystic themes calls into question the durability (in terms of centuries not decades) of the creative *Star Wars* mythology. According to Campbell, new creative mythologies must address the metaphysical and mystical prospect by resonating with age-old issues of spirit, spirituality, and transcendence. A reviewer of the final *Star Wars* film (*Episode III: ROTS*) noted that with regard to mythology, "the theology (or cosmology) isn't developed" (Kondracke, 2005).

Anyone familiar with *Star Wars* would easily recognize the popular saying "May the Force be with you." Although few casual fans would actually be able to explain what the Force *is*, the attraction of the Force as a construct seemed to result from its unnamed simplicity. The Force was about equality; characteristics of the faith include that it is nondenominational, environmental, and inclusive. The Force became such a powerful concept that Jedi was claimed by many American, British, and Australian citizens as a religion on a variety of census surveys during the early years of this decade.

Why, it might be asked, did such a popular concept suddenly seem to lose its attraction? Lucas' most recent prequel dampened interest in this unorthodox religion.[6] *Christianity Today*'s "The Force Is NOT with Them" (2002) documents the decrease in interest. A disgruntled fan base may be one sign that characterizes a loss of mythic power. It stands to reason that Lucas might have lost some mythic appeal owing to a poor treatment of his thematic (or, in fact, an overly detailed and quantified treatment of a transcendental power that required no elaboration). Walt Fisher might contend that the decline in interest resulted from a lack of narrative fidelity and/or cohesion; in other words, the story falls apart either within itself or as a result of comparison with other stories.

Other authors in this anthology propose that new mythologies must somehow address classic themes (e.g., individuality, collectivism, cosmology, metaphysicality) while incorporating new topoi. Whitt, for example, argues that the cybernetic theme can easily be framed within the classic Frankenstein myth and that new mythologies must be prepared to address new modes and forms of life. It is my belief that Lucas trips on his own shoelaces upon the intersection of two themes—the metaphysical prospect and genetics.

## Return of the King

The polesemy of meaning within the title of this section is intentional, as it calls to mind allusions ranging from Jesus to J. R. R. Tolkien. It should be noted that the tales both of Jesus and of Gandalf in Tolkien's mythology provide exemplary treatments of mythic structures that characterize an enduring tale. While Jesus is clearly the Messiah within Christian tradition, his humility becomes one of the most important characteristics of his story. Gandalf embodies the mystic theme of spiritual rebirth with the loss of Gandalf the Grey and transcendence of Gandalf the White. Had Lucas delved more deeply into mythic structures (fairly contemporary tales), he might have made different decisions when attempting to relate the story of Anakin Skywalker. Specifically, by attempting to *geneticize* the Force, Lucas treads on historically shaky territory with regard to myth and mystical structures.

By most accounts, the Protestant Revolution in Europe occurred for several reasons, but one philosophical motive cited by scholars is the phenomenon of humanism and the belief that all citizens are of equal value. It has been noted that the elite clergy during the sixteenth century often evoked a nearly genetic argument regarding papacy, theology, and the priesthood; kings would often call upon the mythic theme of divine grace with regard to their right to rule. While arguments of genetic superiority can be found through modern human history, this theme is often conjured at the folly of the magician. Rarely does the argument of genetic superiority meet with a fond and resonant audience. In fact, elitism and hubris are often dramatic foils that both confront and undo the protagonist; some heroes fail to recover from these stumbling blocks.

Both miscegenation and segregation laws often have at their roots arguments regarding genetic superiority and inferiority. Most modern societies reject discrimination on the basis of genetics, yet there are noteworthy examples of widely accepted appeals to genetic preference in several cultures. Perhaps the most well-known example of contended genetic supremacy is found in the *Herrenrasse*, or master race, that characterized Nazism. Such arguments usually conclude that an individual or group has the right to rule as an inheritance of genetics.

Oddly, the all-inclusiveness of mysticism in the first trilogy of Lucas' work[7] (*ANH, ESB, ROTJ*) seems to have imploded with the subsequent trilogy (*TPM, AOTC, ROTS*) and the introduction of a "midi-chlorian" test (Brooks, 1999, p. 151). With this installment to the *Star Wars* mythology, Lucas inexorably links genetics (via a blood sample) and spirituality (the Force). The result of this synthesis suggests that those who are able to commune with a transcendental power have inherited their ability through birthright.

At this point it is important to note (from a semiotic perspective) that many of the visual symbols used in the *Star Wars* films implicitly critique the epoch influenced by the Nationalsozialistische Deutsche Arbeiterpartei (or Nazi Party). Lucas labeled the minions of evil "stormtroopers." Vader's helmet resembles the German *stahlhelm*. Imperial officers wear outfits similar to those of Nazi "protection squads." Lucas chose these costumes consciously as the embodiment of evil. Therefore, it is curious that genetic inheritance of spiritual privilege (a.k.a. the midi-chlorian birthright of Jedi) would be manifested by an artist (Lucas) who so completely refutes symbols of Nazism.

From a mythic perspective, the idea of divine grace and inherited status has little benefit within most mythologies. Those who claim to inherit their status as a birthright from the gods are often tested by fate, confronted with enormous obstacles, and rarely accepted by their counterparts. Perhaps the most troublesome consequence of the Lucas mythology is the link between power, spirituality, and genetics—links that bode dire consequences in contemporary human history. Within the most recent installment of the Lucas mythology (*TPM, AOTC, ROTS*), ability to use the Force is a birthright; power is the result of an ability to shape the Force; therefore, power is a birthright. This spiritual syllogism empowers only those of proper breeding while sealing the doom of those who failed to inherit the proper chromosomes. Mythology is filled with avatars, deities, demigods, and heroes, but it is rare that birthright alone determines the ascendancy or condemnation of all.

To make the argument more pragmatic, let's assume for one moment that you are Han Solo, a character who mocks the Force and assumedly has a low midi-chlorian count. What happens upon your death? In most mythological traditions, ascendancy would be determined by righteousness, and a fate in the underworld would await those who have been malicious. Regardless of "the Captain's" innumerable good deeds, by *geneticizing* "the Force," George Lucas has doomed Han Solo, who simply dies—there doesn't seem to be an afterlife waiting for him (owing entirely to poor genetics). Consequently, his afterlife is an inheritance instead of a right of deed.

This imperfection in the expanded Lucas mythology seems to be a type of spiritual absolutism. One is either born with the Force or without. One is either born with an afterlife or not. The result is a portrait of the Jedi faith as "an unelected, self-selected, elite band of wise creatures who always have the Force with them, let their feelings tell them what's right and, just in case, have mighty light sabers to whack the bad guys" (Kondracke, 2005, p. 39). It is ironic that such absolutism has been critiqued by Lucas himself, and allegedly the present political climate has been mocked by Lucas (particularly in a characterization of Darth Vader as President Bush [or Vice President Cheney]) for such reductionism.

In the end, Lucas should have been more careful when crafting the tale of a tragic hero. Anakin Skywalker fits the mold of a classic tragic hero because (in the most general terms) he is handed everything, guided to a position of status, and squanders his talent by joining the wrong side. However, in an attempt to enable the narrative of this tragic hero, Lucas rips the fabric of the Force and has developed a poorly articulated myth with regard to spirituality and biology. Genetic right does not sit well as a reason for ascendancy to power. Unwittingly, Lucas actually undercuts the appeal of any character who uses the Force within a *Star Wars* galaxy.

Even though Lucas critiques absolutes (you are either American or not; you are either with me or my enemy), he falls victim to mystical absolutism at the conclusion of his double trilogy. One might ask, what are the consequences of spiritual absolutism? What happens when it becomes impossible for people of various faiths to dialogue about a variety of topics? Perhaps the best way to illustrate the problem that Lucas encounters as he mixes genetics with the mystical prospect is through negative definition.

Imagine for a moment that Lucas wrote a second script—one that is the antithesis of Lucas' published plotline—and that a more promising social message unfolds. Anakin Skywalker is tested for Force ability (on the basis of his mentor's belief that Anakin is very skilled with regard to the Force); now consider the alternative to the most recent trilogy—imagine that Anakin has a midi-chlorian level of zero. In other words, he can use the Force *despite* the fact that his genetic makeup should preclude him from harnessing this transcendental power. Suddenly, the teachings of the Jedi (complete with genetic bias) are questioned. Additionally, within the *Star Wars* universe, segregation on the basis of genetics (along with the policy of teaching only those who prove their genetic predisposition) is challenged. Ben Kenobi—angered that a genetic inferior has been placed under his tutelage—views Anakin with distaste and skepticism. Unable to believe that someone lacking in genetic ability could be the "hero of prophecy," the Jedi council refuses to grant Anakin equal status and all accorded rights. Ultimately, frustrated by his marginalization, Anakin turns toward the only person who will embrace his ability—an emperor of evil. To the dismay of an entire galaxy, the fate of all people is sundered owing to an ancient and avoidable sin—prejudice. Such an alternative to the Lucas plotline would resonate well with egalitarian and emancipatory themes embodied in other science fiction texts, including *Firefly, Blade Runner, Gattaca,* and *The Island.*

It becomes the responsibility of Anakin's heirs (in my "fantasy script" of conjecture) to rebuild the Jedi council, account for the mistakes made through policies of discrimination, and create a just society born from the ashes of ignorance. In this parallel world (placed for the sake of argument in the *Star Wars* universe), Anakin's downfall is the result of prejudice; his salvation—and the salvation of the Core Worlds—is recognition and acceptance. Ultimately, if we attempt to maintain the narrative thread woven by Lucas, how fitting that his redemption would come at the hands of Luke Skywalker. It is assumed (in my alternative world of conjecture) that Luke also lacks a genetic predisposition toward Force sensitivity; yet somehow Luke has also become a Jedi. This time, however, Anakin's doubters (e.g., Ben and Yoda) rally around Anakin's son and ultimately atone for past prejudice by mentoring and supporting young Skywalker.

Let us also analyze the antithesis represented by other popular texts (i.e., by contemporaries of Lucas) that deal effectively with issues of genetics. J. K. Rowling in her *Harry Potter* series uses a Muggle-born character to dismiss the idea that only a certain breed of humanity can harness magic. Try to envision Hogwarts as a school for only those of proper breeding and lineage. Imagine a world called Middle-Earth, in which only creatures of a certain physical stature are able to accomplish particular deeds. In such a world, Frodo would hand the Ring of Power to Gandalf and his journey would be over. Simply put, other new mythologies (e. g., Guillermo del Toro's *Pan's Labyrinth* [*El Laberinto del Fauno*]) have given more careful consideration to the issue of genetics, and the resulting texts are tremendously rich and resonate with culture. The treatment of Lucas, on the other hand, seems implicitly to embrace absolutism and (most likely by accident) places genetic inheritance upon the highest of pedestals.

## Conclusion

It is surprising that Lucas decided to address the mystical prospect in what must be labeled a reductionist manner considering how few mythic texts intentionally make a link between salvation and genetics. Such links, when they are made, are usually found in polytheistic traditions and are the result of nature-driven mythos. Lucas, on the other hand, grounds the identity of Anakin Skywalker within the tale of a messiah—a conventional theme in the monotheistic Christian tradition. Like the Christian messiah, Anakin is born of mystical origins and inherits the ability to perform miracles. All comparisons to a Christian messiah stop here. Anakin is filled with hubris; he believes that his genetic powers will save the galaxy. Additionally, the Christian messiah does not close the doors of salvation for those born of a certain heritage. In fact, the Christian faith is relatively inclusive within the context of modern religion (allowing one to be born again). Anakin's tale, told through Lucas, suggests that biology is the precursor to ascendancy and salvation vis-à-vis the Force.

As established earlier in the epoché of this project, I entered into my critical study of the Lucas mythology as a recalcitrant fan. Admittedly, my reading of the *Star Wars* universe is personal; consequently, I am open to criticism—in fact, I embrace the potential dialogue with colleagues. However, I adamantly contend that the most recent *Star Wars* trilogy has met with less enthusiasm (than the first trilogy) owing largely to the mythic missteps of George Lucas. In particular, Lucas paints himself into a corner with his poor conception of a synthesis between biology and theology. The result is a theocracy, known as the Jedi Order, laden with implications that evoke modern-day Aryanism.

In the end, the *Star Wars* universe and the tale of Anakin Skywalker will be judged by the test of time. Although Lucas should be commended for a generally effective treatment of several mythic themes (including collectivism, individualism, and cosmology), his immature articulation of the mystic prospect may undermine an otherwise noteworthy creative myth (measured, again, by centuries rather than decades). Ultimately, the Achilles heel in this mythology yields to others (both authors and myths) as more accurate representations of what could be construed as New Mythology. Unless Lucas can more fully articulate his vision of metaphysicality and spirituality (perhaps with future films), his final trilogy (*TPM, AOTC, ROTS*) will close the mythological *Star Wars* universe not with a big bang but with a rather dull thud.[8]

## Notes

1 The reference to Lucas as "the Beard" is taken from http://www.imaginarycinema.com/lucas.html

2 *Epoché* (EH-pah-kay) is a Greek term describing a state in which consciousness is subject to introspection and critique. Conceived by Aristotle, the *epoché* is a pillar within skeptical inquiry.

3 Meaning is within both the writer and the reader; therefore the intention of Tolkien is not more significant than the import placed on his cosmology by the audience.

4 This is the implication one can derive from *ANH, ESB,* and *ROTJ*. However, *TPM, AOTC* and *ROTS* muddy the waters somewhat.

5 According to Campbell (in Campbell & Moyers, 1988, p. 56), "A metaphor is an image that suggests something else. The reference of the metaphor in religious traditions is to something transcendent that is not literally anything."

6 The September 24, 2006 episode of *Filter* on G4 TV began its "From Cool to Crap" top-20 list with "*Star Wars* movies."

7 It is worth noting that the vague and ephemeral treatment of the Force found in Lucas' early work is actually consistent with the musings of Joseph Campbell cited earlier in this chapter.

8 At the time of writing we are preparing to celebrate the thirtieth anniversary of *Star Wars: A New Hope*. I will stand corrected regarding the analysis of *The Phantom Menace, Attack of the Clones,* and *Revenge of the Sith* if we equally celebrate these three movies in 2029.

# References


Altman, I., Vinsel, A., & Brown, B. (1981). Dialectic conceptualizations in social psychology: An application to social penetration theory and privacy regulation. In L. Berkowitz (Ed.), *Advances in experimental psychology* (Vol. 14, pp. 107–160). New York: Academic Press.

Bakhtin, M. (1990). *Art and answerability*. Austin: University of Texas Press.

Bakhtin, M. (1994). *The dialogic imagination,* 9th edition. Austin: University of Texas Press.

Baxter, L. (1994a). A dialogic approach to relationship maintenance. In D. J. Canary & L. Stafford (Eds.), *Communication and relational maintenance* (pp. 233–254). San Diego, CA: Academic Press, Inc.

Baxter, L. (1994b). Thinking dialogically about relationships. In R. L. Conville (Ed.), *Uses of structure in communication studies* (pp. 23–37). Westport, CT: Praeger Publishers.

Brooks, T. (1999). Star *wars: Episode I: The phantom menace.* New York: Ballantine.

Buber, M. (1970). *I and thou: A new translation,* with prologue and notes by Walter Kaufman. New York: Touchstone.

Campbell, J. (1991). *The masks of God: Occidental mythology*. New York: Penguin Press.

Campbell, J. & Moyers, B. (1988). *The power of myth*. New York: Doubleday.

Fisher, W. (1987). *Human communication as narration: Toward a philosophy of reason, value, and action*. Columbia: University of South Carolina Press.

Foss, S. K. (2004). *Rhetorical criticism*. Long Grove, IL: Waveland.

Kondracke, M. (2005). Tales from the Dark Side don't live up to the hype. *Chicago Sun Times,* June 5, 2005, p. 39.

Lehrman, D. (1953). A critique of Konrad Lorenz' theory of instinctive behavior. *The Quarterly Review of Biology, 28 (4),* 337–363.

Leinenweber, C. (1977). Socialists in the street: The New York City Socialist Party in working class neighborhoods, 1908–1918. *Science and Society, 41,* 152–171.

Lorenz, K. (1937). Ueber den Begriff der Instinkthandlung. In D. Lehrman's (1953), A critique of Konrad Lorenz' theory of instinctive behavior, *The Quarterly Review of Biology, 28 (4),* 337–363.

Maines, D. (1989). Further dialectics: Strangers, friends, and historical transformations. In J. Anderson (Ed.), *Communication Yearbook, 12,* 190–202. Newbury Park, CA: Sage.

Montgomery, B. (1993). Relationship maintenance versus relationship change: A dialectical dilemma. *Journal of Social and Personal Relationships, 10,* 205–223.

Rawlins, W. (1989). A dialectical analysis of the tensions, functions, and strategic challenges of communication in young adult friendships. In J. Anderson (Ed.), *Communication Yearbook, 12*, 157–189. Newbury Park, CA: Sage.

Rawlins, W. (1992). *Friendship matters: Communication, dialectics, and the life course*. Hawthorne, NY: Aldine De Gruyter, Inc.

Rychlak, J. (1976). The multiple meanings of dialectic. In J. F. Rychlak (Ed.), *Dialectic: Humanistic rationale for behavior development* (pp. 1–17). New York: Karger.

Shelton, A. (2006). Planchette, My Love. *Cultural Studies/Critical Methodologies, 6* (3), 311–329.

The Force is not with them. (2002). *Christianity Today, 46* (12), 22.

Werner, C., & Baxter, L. (1994). Temporal qualities of relationships: Organismic, transactional, and dialectical views. In M. Knapp & G. Miller (Eds.), *Handbook of interpersonal communication,* 2nd edition (pp. 323–379). Newbury Park, CA: Sage.

• CHAPTER THREE •

# Booyahs, Sonic Cannons, and a 50,000-Watt Power Cell: *Teen Titans'* Cyborg and the Frankenstein Myth

*David Whitt*[*]

On September 15, 2006, two mainstream journalistic outlets engaged in an intriguing bout of online one-upmanship. At 7:28 a.m. EST *Yahoo! News* posted a story from the Associated Press titled "Man's bionic arm provides hope for GIs" about fifty-nine-year-old Jesse Sullivan, a former utility lineman who lost both arms during an accident in 2001 and became the first amputee with a thought-controlled artificial arm. Minutes later, at 7:52 a.m. EST *USAToday.com* updated their Tech section with "Meet the $4 million woman," the inspiring story of twenty-six-year-old Claudia Mitchell, a former U.S. Marine who lost her left arm during a motorcycle accident in 2004. After reading about Sullivan's bionic arm in *Popular Science,* Mitchell was determined to receive her own bionic arm and was fitted with one even more technologically advanced than Sullivan's, allowing her to feel heat and pressure. Fusing the organic with the artificial involves the complex and delicate procedure of "muscle reinnervation," which for Mitchell involved grafting shoulder nerves to the pectoral muscle. Movement occurs when computer electrodes within the prosthetic arm receive electrical impulses from the brain, translating thought into action.

* Dr. David Whitt (M.A., University of Nebraska; Ph.D., University of Nebraska) is an Assistant Professor of Communication at Nebraska Wesleyan University.

Describing how this technology has changed her life Mitchell exclaims, "I can open a spaghetti jar and hold it up at an angle and use a spoon to empty it out. Small things like that may seem trivial to a two-armed person, but it is very exciting to me" (Davis, 2006, p. 2).

The remarkable success of Sullivan's and Mitchell's bionic arms has inspired the Defense Advanced Research Projects Agency (DARPA), the central research and development branch of the Department of Defense. In April 2006 DARPA announced it would invest $55 million into a prosthetic arm project headed by Johns Hopkins University's Applied Physics Laboratory. One of the dozens of institutions contributing to this project is the University of Utah, responsible for developing a "peripheral nerve interface" that will relay nerve impulses from an amputated limb to a bionic arm. According to University of Utah bioengineer Greg Clark, "Imagine an artificial arm that moves naturally in response to your thoughts, that allows you to feel both the outside world and your own movements, and that is as strong and graceful as an intact biological arm. That's what our researchers, teaming with others around the world, are setting out to achieve" (Britt, 2006, p. 1).

However, DARPA is not alone in its quest to merge technology and human physiology. Hughes (2004) describes how cochlear implants have made it possible for the deaf to hear, and electronic vision systems are enabling the blind to see. Such remarkable twenty-first century advancements in prosthetics, artificial implants, and computer systems are creating a new kind of human, first defined by author Manfred Clynes back in 1960 as a "cyborg" (Halacy, 1965).

Dumit and Davis-Floyd (1998) explain, "Cyborgs are symbiotic fusions of organic life and technological systems" (p. 1). Once thought to exist exclusively within the realm of science fiction, the cyborg is now a dominant and ubiquitous presence in contemporary culture. Consequently, Gray, Mentor and Figueroa-Sarriera (1995) argue that we now live in a "cyborg society" as machines "are intimately interfaced with humans on almost every level of existence not only in the West and Japan but among the elite in every country of the world" (p. 3). The applications of this machine/human merger are limited only by the imagination, as researchers are experimenting with everything from implanted medical data microchips, to various artificial organs, to a cell phone that can be installed in your teeth. However, as our transformation from the biological to the cybernetic continues many questions are raised as to how this will impact our individuality and humanity.

Such philosophical concerns reflect what Clark (2003) refers to as a "double whammy" within the classic cyborg image. He explains, "First, we care about the potential of technology to become integrated so deeply and fluidly with our existing biological capacities and characteristics that we feel no boundary between ourselves and the nonbiological elements. Second, we care about the potential of such human-machine symbiosis to transform (for better or for worse) our lives, projects and capacities" (p. 24). While scientists and scholars continue to debate the moral and ethical implications of merging machine and body, another set of visionaries, science fiction television and film writers, have used their imaginative talents to create a unique gallery of cyborg characters. According to Holland (1995) cyborg films such as *Terminator, RoboCop,* and *Universal Soldier* "foreground questions of dualism and personal identity especially clearly, and highlight contemporary concerns about the effects of technology on the human 'self' in the present and the future" (p. 157). Certainly Holland's statement extends to popular cyborg television shows like the 1970s *The Six-Million Dollar Man* and *The Bionic Woman* (and its 2007 reboot), and more contemporary programs such as *Jake 2.0* and *Battlestar Galactica*.

However, cyborg motion pictures and television are more than just fictional action-adventures; they are powerful mythic texts that comment on contemporary culture within a modern or futuristic setting. In terms of cyborg science fiction the foundational myth of this genre is Mary Shelley's (1994) *Frankenstein*. Gray, Mentor, and Figueroa-Sarriera (1995) state, "Many see Mary Shelley's monster, Frankenstein's creature, as the first cyborg" (p. 5). They explain that stories such as *Frankenstein* "raise important questions about our relationship to the technological systems we unleash on the world" (p. 6). Thus, examining *Frankenstein* helps to frame how this myth addresses our hopes and fears regarding cyborg development. To better understand how a nineteenth-century novel can influence twenty-first-century thought I analyze the Frankenstein myth in relation to the African American character Cyborg on the Cartoon Network series *Teen Titans*. However, before this can be done I must first discuss *Frankenstein* and its value as a mythic text.

## The Frankenstein Myth

Published in 1818 by Mary Wollstonecraft Shelley, *Frankenstein, or, The Modern Prometheus* is recognized as a classic work of English Gothic literature. The novel tells the story of Victor Frankenstein, a student of the natural sciences who becomes obsessed with the possibility of "bestowing animation on lifeless matter." After years of research he discovers "the cause and generation of life," and realizing the great power he has achieved, labors to build an eight-foot-tall human being from body parts stolen from dissecting rooms and slaughterhouses. Two years later Frankenstein succeeds in his quest, but on seeing his creation awaken, is horrified by the result. Frankenstein flees from his lab, and his creature disappears into the woods and begins the long process of self-discovery. Acquiring language through listening to others and reading Milton's *Paradise Lost* and Plutarch, the creature begins to question his existence. He states,

> As I read, however, I applied much personally to my own feelings and condition. I found myself similar yet at the same time strangely unlike to the beings concerning whom I read and to whose conversation I was a listener. I sympathized with and partly understood them, but I was unformed in mind; I was dependent on none and related to none. 'The path of my departure was free,' and there was none to lament my annihilation. My person was hideous and my stature gigantic. What did this mean? Who was I? What was I? Whence did I come? What was my destination? These questions continually recurred, but I was unable to solve them. (p. 123)

Angry at the fear, injustice, and violence he has experienced from humanity, the creature asks Frankenstein to make him a female companion. Victor initially agrees to this but then later destroys the work he has already completed. The creature vows revenge for this action and Victor's loved ones are killed one by one. The novel's denouement describes Frankenstein's arduous quest to destroy the monster, following him to the cold and icy wastelands of the north. However, neither creator nor creation finds satisfaction in the end. Frankenstein dies on a ship, unsuccessful in destroying his "daemon." The monster's motivation for living now gone, he laments his future and disappears to die alone.

Almost two centuries after its publication *Frankenstein* is still recognized as a work of remarkable depth and insight into the human condition. Its themes of science, ethics, and individuality are timeless, as relevant in the new millennium as they were in the nineteenth-century. Slusser (1992) argues that Shelley's work "is indeed the first SF novel...it seems to be the first work in which the processes of traditional fiction and modern science meet in any meaningful fashion" (p. 46). O'Mahony (2002) adopts a more apocalyptic perspective of the novel, stating that *Frankenstein* "set the agenda for science as the harbinger of man's destruction" (p. 6).

But beyond its distinction as the literary "big bang" for the science fiction genre, *Frankenstein* is also an important mythic text. Lykke and Braidotti (1996) state that the novel "has been more or less canonized as *the* (author's italics) myth of modern science monsters" and "touches strongly on these feelings of fear and aversion to the non-human/human boundary figure" (p. 15). According to Bloom (1994) *Frankenstein* is significant because it "contains one of the most vivid versions we have of the Romantic mythology of the self" (p. 215). A response to neoclassicism and the Industrial Revolution, the Romantic movement celebrated emotionalism and individuality as reflected in the works of Rousseau, the Brontë sisters, Keats, and Wordsworth.

While Shelley's contributions to this movement included other novels and poems, *Frankenstein* is unique for its dark Gothic tone, combining horror, emotion, and individual angst. "The profound dejection endemic in Mary Shelley's novel," Bloom explains, "is fundamental to the Romantic mythology of the self, for all Romantic horrors are diseases of excessive consciousness, of the self unable to bear the self" (p. 221). This question of self is central in the novel as Frankenstein's conscience is tortured by the morality of his Promethean act, while the creature struggles to rationalize not only who he is, but also his purpose. In fact, Bloom argues that the monster is paradoxical because he is *more human* (author's italics) than his creator since he is "more lovable than his creator and more hateful, more to be pitied and more to be feared, and above all more able to give the attentive reader the shock of added consciousness in which aesthetic recognition compels a heightened realization of the self" (p. 215). As a product of humanity's faith in science and technology the creature's journey of self-discovery transcends Romanticism, making him the embodiment of our contemporary digital culture reflected through the image of the man/machine cyborg.

Wagner and Lundeen (1998) contend that Shelley's novel is "one of the great myths of modern times, and like all great myths it lends itself to a variety of revisionings and adaptations" (p. 52). Frankenstein mythology has been applied, for better or for worse, in a variety of contexts. Campbell, Martin, and Fabos (2007) explain that Shelley's novel has been transformed into various forms including the 1931 film version starring Boris Karloff, the 1960s television comedy *The Munsters*, and even the breakfast cereal Frankenberry. They argue that in these recycled forms "Shelley's powerful themes about abusing science and judging people on the basis of appearances are often lost or trivialized" (p. 22). However, when applied to more modern texts, like cyborg science fiction, these themes remain both prophetic and socially relevant. For example, Rushing and Frentz (1989) examine the Frankenstein myth in relation to cultural fears about technology through the 1980s films *Rocky IV, Blade Runner,* and *The Terminator.* Other Frankenstein-inspired characters such as Darth Vader, RoboCop, and The Borg from television's *Star Trek,* made the cyborg a popular culture fixture, while more recent films such as *The Matrix* trilogy imagine this creation for our digital age. Arguably, most cyborg television and film texts are geared toward teenagers and adults, with mind-bending time travel narratives, sexuality, and graphic violence. However, in the new millennium, technology-embodied characters are appealing to a younger demographic. One in particular, the teenage superhero Cyborg from the television cartoon *Teen Titans,* has put a unique spin on the Frankenstein myth.

## Cyborg's Origins

Debuting on the Cartoon Network in 2003, *Teen Titans* is an animated version of the DC Comics title of the same name first published back in the late 1960s (Cadigan, 2005). The Titans are a group of teenage superheroes, each with his or her own extraordinary abilities, who, not surprisingly, fight crime and injustice. Arguably, the most popular version of the comic book *Teen Titans* was its 1980 relaunch lineup, which included Robin, Starfire, Raven, Kid Flash, Wonder Girl, and Cyborg. Although the focus of this analysis is on the *Teen Titans* cartoon character Cyborg, it is important to first discuss his origin as it was first told within the pages of DC Comics.

Issue one of the four-issue miniseries *Tales of the New Teen Titans* (1982) tells the story of how teenager Victor Stone was transformed into Cyborg. Victor's parents, university research scientists, searched for ways to boost their child's intelligence. By the time Victor reached puberty he was a multisport athlete and had developed a high I.Q. One day while visiting his parents at their lab Victor discovers that a creature from another dimension has killed his mother. Before his father manages to send the creature back to its dimension the creature attacks Victor, burning his flesh and dissolving his bones. To save Victor's life his father uses a cybernetic suit he has developed for the military and surgically implants various components into his son's deformed body. As Victor himself explains,

> Bone was replaced with steel. Flesh with special polymers. Veins and arteries were encased in plastic tubings, and metal straps bound it all together. Piece by piece I was rebuilt. A modern Prometheus, my father thought. A latter day Frankenstein was more like it. (Wolfman and Perez, 1982, p. 15)

On regaining consciousness Victor sees that most of his original body has been replaced by artificial components. His arms and legs are now prosthetics, most of his torso is covered with a silver alloy, and the left side of his face is encased in a metallic shell. Victor's father explains that the cybernetic implants should make him better than ever; however, Victor is shocked at his new appearance and cries, "Why couldn't you let me die!?!?"

After several months of rehabilitation Victor learns to adapt to his new cybernetic body. Although Victor now has superior strength, speed, hearing, and various weapons and electronic devices that can attach to his right arm, his appearance generates public distrust. During a walk in the park he explains,

> I felt their eyes boring into me. Fear. Hate. It hurt so much. But never so much as when kids ran away from me. I was a monster. A blasted Frankenstein, And I cursed my father all over again for not letting me die. (Wolfman and Perez, 1982, p. 19)

Eventually, Victor discovers that he can use his remarkable abilities for good, and, with the help of the other Teen Titans, learns to forgive his father.

The use of Frankenstein mythology within this special issue of *Teen Titans* is rather obvious. Besides the specific references to Frankenstein and Prometheus within the text, the name Victor is, no doubt, a nod to Shelley's protagonist. Additionally, like Frankenstein, Victor's father creates life from lifelessness and in the process, alienates his creation, in this case, his own son. Victor is also a teenager and an African American, which places him within both a social and cultural minority. Together, his now less than human appearance, race, and age make Victor the ultimate outsider, an individual seeking acceptance from society, much like Frankenstein's creature.

In dramatic contrast to the DC Comics *Teen Titans* miniseries, Cyborg's origin in the Cartoon Network's animated series is more ambiguous. Similar to his comic book counterpart the small-screen Cyborg has prosthetics, giving him strength, speed, and sight. Most of his body and the left side of his face are covered with titanium armor, and he can access a variety of tools from his arms, including a sonic cannon and a blowtorch. However, despite their similar physical appearance there are some minor differences between the two. In the television series he is called Cy or Cyborg rather than Victor, and the details of how he came to be are vague. The fifth season episode "Go" gives some clues to Cyborg's transformation when he tells the shape-shifting teenage superhero Beast Boy, "I had an accident and now I'm a monster, alright? A cyborg." There is no flashback sequence to the tragic event, no mention of his father's experimentation, no rehabilitation montage, but the bitterness Cyborg feels toward his transformation is still evident in his Frankenstein-inspired self-referential label as a "monster."

*Teen Titans* is rated Y7 for fantasy violence, so perhaps the animated retelling of Cyborg's transformation, so graphic in the comic book, would not be appropriate for younger audiences. Therefore, the lack of any scenes detailing Cyborg's origin is not surprising. However, what is more important is not how Cyborg came to be but how he learns to accept who he is and what he has become. While the Cyborg-centric episodes of *Teen Titans* typically focus on his technological abilities, it is his internal conflict that makes Cyborg more than just a two-dimensional cartoon character. Several episodes highlight Cyborg's struggle for humanity while drawing upon themes within the Frankenstein myth.

## The Cyborg Equation

The first season episode "The Sum of His Parts" begins in a park where the Titans are relaxing and playing football. Cyborg goes out for a pass, when suddenly his power cell fails. He is reactivated but has only thirty minutes of energy left. He tells Starfire, "It's just a natural part of being unnatural." On his way back home to retrieve a new battery, a young boy sees Cyborg and tells him, "You're my favorite, 'cuz look..." and then he removes the baseball glove on his right hand to show a prosthetic arm and enthusiastically proclaims, "...you're just like me." Cyborg replies, "I guess I am." Shortly thereafter the Titans are called away to do battle with the magical villain the Amazing Mumbo. During the conflict, Cyborg's thirty-minute auxiliary power supply dies and he is suddenly below ground. He wakes up on an operating table and meets a mechanical medic called Fixit, who tells him to rest since his power source is not fully recharged. Surrounding Cyborg are hundreds of smaller robots whom Fixit has repaired. Cyborg tries to leave but Fixit tells him, "You are not repaired. Not fully. My examination revealed many flaws. Serious flaws." Cyborg is alarmed by this claim, but Fixit says there is nothing to worry about and persuades him to stay to finish repairs.

After his power is restored to 100% Cyborg again tries to leave, but Fixit explains,

> Fixit: Your biological components are imperfect and must be replaced.
> Cyborg: What if I don't want to be repaired? What if I like me this way?
> Fixit: Such thoughts come from your imperfect biological components. They will be replaced momentarily.

Cyborg manages to rip away the robe Fixit is wearing and is stunned to see biological components combined with machinery; he is also a cybernetic being. Ignoring this revelation Fixit returns to his singular goal of repairing Cyborg:

> Fixit: The biological malfunctions are growing worse. We must begin at once. It is for the best.
> Cyborg: No it isn't. If you take out my biological components you take out the best part of me, the part that makes me who I am.
> Fixit: Your memories, your self will be preserved. I will download them into your new brain.

Cyborg resists and after a brief scuffle defeats Fixit, who then comes to have a renewed appreciation for his human senses:

> Fixit: So beautiful. I had forgotten how beautiful...the world...through your eyes.
> Cyborg: Through human eyes, the same kind you have.
> Fixit: Perhaps I am the one in need of repairs.
> Cyborg: Then maybe I'm the guy who can help repair you.

The episode's final scene shows the young boy from the beginning who sees Cyborg again and says to his friend, "See, he's just like me" and shows his prosthetic hand again. Cyborg tells the boy, "I am just like you, but it's not your arm that makes us the same, it's the stuff connected to it."

"The Sum of His Parts" highlights an ideological struggle between two artificially enhanced beings: Cyborg, who fights to keep his humanity and Fixit It, who strives to suppress his own. While Fixit views Cyborg's biological components as "imperfect," Cyborg sees them as part of what makes him a unique individual. Much as in the prologue and epilogue scenes with the young boy and his prosthetic arm, Fixit becomes comfortable in his own cybernetic skin and human emotions. These themes of acceptance and understanding are common to other Cyborg-centric episodes.

In the opening of "Only Human" the Titans are seen working out in their training room. Cyborg is trying to sustain a large amount of weight but struggling. The power monitor on his arm reads 100% (maxed out), but despite encouragement from Robin, Cyborg quits. Robin tells Cyborg that all he wanted him to do was kick it up another notch, to which Cyborg replies,

> I don't have another notch. I'm not like you. When I say I can't, I can't. When I was an athlete, when I was human, I loved pushing my limits. Getting stronger, faster, better, just trying harder than I ever had before. My coaches would always tell me to give 110%, and I always did. My muscles are mechanical now, limits are built in. No matter how hard I try, 100% is all I've got.

Cyborg retreats to the Titans game room to unwind. While playing an interactive Internet videogame Cyborg defeats his online opponent called Atlas (a name from Greek mythology), who demands a rematch. Suddenly, Atlas, who is actually a ten-foot-tall robot, materializes out of nowhere and begins fighting the Titans. During the long battle, Cyborg's power cell falls to 50%; Atlas captures the Titans and challenges Cyborg to another battle at an abandoned stadium. After recharging, Cyborg goes to the stadium, and in a mocking tone Atlas says,

> Atlas: I did not think you would come. Humans scare so easily.
> Cyborg: I'm only half human.
> Atlas: Then half of nothing is still nothing.

Their battle resumes and soon Cyborg's energy output reaches its limit of 100% and he is defeated again. Leaving the stadium, without his teammates, a dejected Cyborg is next seen walking down the street, wondering aloud whether he should give up. Cyborg's reflection in a store window begins talking back to him, serving as his conscience:

> Cyborg: What else can I do? I gave everything I had. 100%
> Reflection: Give more! Your friends are in trouble. You have to go back. You have to win.
> Cyborg: I can't win. Atlas is stronger. Atlas is faster. He's just a better robot.
> Reflection: But you aren't all robot. Half of you is human, and that's the half that can beat him.

Cyborg then goes back and fights Atlas for a final time. As before, Cyborg's power eventually maxes out at 100%. Atlas tells Cyborg that he has reached his limit and cannot win. Struggling, Cyborg says he can win, and his power jumps from 110% to 120% and finally 130% before shattering the energy output monitor on his arm. Cyborg then throws Atlas to the ground.

> Atlas: No more. I am defeated. You are a better robot.
> Cyborg: No. I'm a better person.

The captured Titans are now free, but Robin notices Cyborg's energy level monitor is broken and suggests he gets a new one. Cyborg replies, "Don't need it anymore. A body may have its limitations, but if I put my mind to it there's no limit to what I can do. After all, I'm only human."

Cyborg's crisis of confidence in "Only Human" is the first time the Titan doubts his ability to defeat a foe. Eventually, Cyborg finds the strength to defeat Atlas not from an internal energy source, but from his own desire. Cyborg's human half is ultimately more powerful than his cybernetic half, reminding him that despite appearances he is still a living, breathing, and feeling person. This focus on Cyborg's character, driven by personal resolve rather than computer programming, is further explored in "Titans East."

In this two-part episode Cyborg battles his arch nemesis Brother Blood, leader of the terrorist organization the H.I.V.E. Academy (The Hierarchy of International Vengeance and Extermination). Blood is a formidable and dangerous enemy, with enhanced strength, agility, and mental abilities. Cyborg was once undercover in H.I.V.E and

Blood has had a personal vendetta against him ever since. However, what Cyborg does not realize is that when he first hacked into the H.I.V.E. mainframe computer Blood reversed the process and downloaded the Titan's blueprints. Blood has now transformed himself into a cybernetic being similar to Cyborg, threatening to use his new implants to enhance his mental powers to control thousands of students.

On a more personal level, Blood is also obsessed with what lies within Cyborg's circuitry that has allowed the Titan to resist his power. During a fierce battle between the two, Blood systematically dismembers Cyborg, first cutting off his right hand, then punching a hole in his chest, slicing off his left arm, and finally both legs. Blood then tells Cyborg he can "make a man" out of him and reaches into his open chest asking,

> Blood: What is it, Cyborg? I've combed through your blueprints, peeled back your armor, stripped off your machinery, but still you defy me! Where in this vile contraption is the part that allows you to resist?
>
> Cyborg: It's not in the machinery, is it? It's not the machine that resists you. It's me. My spirit. That's the part you can't break. I don't need you to make me a man. I already am one!

On the verge of defeat Cyborg taps into a hidden yet remarkably powerful energy source within his circuitry, miraculously reassembles himself, and with his increased powers defeats Blood.

Blood's offensive against Cyborg is vicious and surgically precise in its execution. Surprisingly, despite losing his appendages, which leaves him completely defenseless, Cyborg refuses to concede defeat. Blood believes Cyborg's strength lies within in his circuitry and not that which makes him human. While cybernetics has allowed Cyborg to live from a biological standpoint, in the end his identity and spirit make him more man than machine. However, Cyborg soon discovers that while being a machine allows him to do remarkable things, it can also have its drawbacks and consequences.

In "Overdrive" Cyborg awakes and accesses his daily schedule on a computer. He is next seen engaging in various activities including playing chess, basketball, and videogames. Although Cyborg has been busy from morning to night, at day's end he wishes he could do more. His eye catches a TV commercial for the Maximum 7, a new high-speed microchip that Cyborg believes can give him the upgrade he desires. On delivery and installation of the Maximum 7 chip, Cyborg and the Titans fight the powerful villain Adonis. With Cyborg's heightened speed and strength Adonis is quickly dispatched, and Beast Boy comments that his teammate is "more cyborgy than ever."

Later that day the Titans encounter Billy Numerous, an old nemesis who can duplicate himself several times over. After an intense battle Numerous escapes, and the Titans, who are exhausted from the engagement, decide to go home and rest before attempting to track him down. However, with his new microchip Cyborg insists he can keep going and decides to draw upon his "Sleep" energy supply to sustain his maximum power. Cyborg eventually finds Numerous and does battle with him several more times, but the Titan is outwitted on each occasion. To give himself the power he needs to continue fighting, Cyborg taps into his stored energy designated for "Hunger," "Fun," and "Emotion," the last of which has an immediate effect upon his demeanor.

The next morning the Titans see Cyborg sitting in a chair connected to several power cables. Cyborg informs the team he no longer needs sleep and has even deleted his sense of humor. Worried about his friend's dramatic change of personality, Beast Boy comments "Cyborg's always had chips for brains, but he is turning into the one thing I never though he could be. A robot." During the Titans' final encounter with Numerous, Cyborg's system overloads and he shuts down. He awakens back at the Titans' headquarters and is told that although the Maximum 7 chip was compatible with his cybernetic implants, when he started shutting down his other programs his system became stressed. The chip is eventually removed and the final scene shows Cyborg sleeping through the night.

## Implications

"Where does humanity end and the machine begin?" This question posed by Wagner and Lundeen (1998) is discussed within the context of our continual cyborg evolution. They argue that the human-machine merger raises important questions about power, responsibility, and even immortality, stating, "It is not hard to see that this possibility, even if it exists only in the realm of narrative, opens new vistas for the mythic imagination" (p. 55). Traditionally, scholars have viewed this vista from the popular culture mountaintop, tending to critique the more commercially successful cyborg characters such as the Terminator, *RoboCop*, and more recently Neo. While the small-screen Cyborg may not achieve the iconic status of his big-screen contemporaries, he certainly has the potential to be as meaningful a subject of study in terms of both mythic analysis and social relevance.

These Cyborg-centric episodes of *Teen Titans* reflect the Frankenstein myth in several ways. First, the series addressed themes inherent in Mary Shelley's novel such as faith in science and the search for self. Cyborg represents a modern interpretation of this myth, a being created not from human organs and appendages but from circuitry and metallic alloys. Like Frankenstein's creature, Cyborg has incredible strength and speed, but this is where the physical comparisons end. Cyborg has weapons and tools built into his body, an internal power cell, and he can interface with computers. Additionally, what distinguishes the two is their purpose of being. Cyborg's transformation from fully human to half machine was necessary to save his life, not the obsessive act of a scientist wanting to play God and create life. In this way, the origin of both individuals influences their attitudes toward existence and their purpose of being.

In terms of Shelley's theme of identity the *Teen Titans* episodes "The Sum of His Parts" and "Only Human" address Cyborg's duality of self. While Frankenstein's creature was built from human body parts, Cyborg was constructed from organic and artificial elements. This combination is potentially damaging to the individual psyche, as the mind must accept the alteration of the body. According to Wilson (1995) the struggle to adjust to technological modification creates within the individual a split consciousness that alters his or her perception of humanity. Wilson contends that in becoming a cyborg we will always remember how we once were and, therefore, may feel "disgust" at our transformation and the "personal deficiencies" (p. 255) that compelled change. This internal conflict impacts the individual on a deeply personal level since part of his or her essence has been replaced or enhanced by technology. Thomas (1995) contends that from this change "technology becomes the determining factor in the definition of the body's physical rearticulation, the material for its sense of performed identity" (p. 38). While Cyborg certainly reflects this tension, unlike Frankenstein's creature he is not a tortured soul full of individual angst and disdain for others. Cyborg's optimism conveys hope that our transformation into cybernetic beings will not be met not with fear but rather with understanding. In this way Cyborg is the antithesis of Frankenstein's creature, comfortable with his appearance, confident in his purpose, and, most importantly, accepted by society.

However, despite his outward demeanor Cyborg has not achieved, to use Hegelian terms, a true balance or synthesis between the organic and the artificial. Cyborg's internal conflict is reflected by his continual effort to maintain his humanity within a cybernetic body superior to flesh and bone, but with imperfections. In other words, Cyborg's greatest enemy is not from a rogues gallery of Titan villains such as Atlas, Adonis, or Brother Blood, but rather the implants that keep him alive and provide him with extraordinary abilities. In each Cyborg-centric episode the Titan's trials and tribulations are caused specifically because of his own technological limitations, which are ultimately overcome by his strength of character and belief in self. In this way the writers of *Teen Titans* offer a more sanguine view of cyborg development, unlike the dystopian perspective more common in popular culture.

While there are thematic parallels between Cyborg and Frankenstein's creature, what is perhaps more intriguing is how a teenage animated character connects on a mythic level with a preteen audience. Given the show's rating of Y7, the target demographic for *Teen Titans* is young children who are likely familiar with Frankenstein from Halloween costumes rather than reading an early nineteenth-century novel. However, as a modern interpretation of the Frankenstein myth, Cyborg is a character to whom children can relate on some level. Despite the fact that the majority of his body covered in cybernetic implants Cyborg is also a teenager who loves junk food, plays videogames, and argues with his friends over the remote control. Moreover, Cyborg is technologically savvy, able to download complex computer programs and build various electronic devices. His attitude and personality are those of a typical twenty-first-century adolescent, reflecting interest, enthusiasm, and inherent trust toward technology. Arguably today's youth, more so than their parents, have embraced or, at the very least, are more comfortable with technology such as computers, the Internet, and videogames. Defined by Howe and Strauss (2000) as the Millennials, this digital generation was born into an era of connectivity, in which Websites are as ubiquitous as fast food (and just as empty), text messaging allows students to cheat on exams, and videogame opponents are online and can live anywhere in the world. The time spent by children and adolescents using technology has naturally generated concern from adults and educators. However, according to Hutchby and Moran-Ellis (2001):

> The important question is not 'what is the impact of technology use on childhood?' (a frequently posed adult concern, to which the answer is often couched in wholly negative terms), but rather, what are the shapes and the outcomes of specific, situated encounters between children and technologies: how do children interact with, and in light of, the affordances that technologies have; how do those affordances constrain such interactions; and how are the complex relations brought about here consequential for our understanding both of children themselves and of technological forms? (p. 3)

In other words, how do children make sense of technology and conceptualize its impact on their own lives? Hutchby and Moran-Ellis' questions reflect Marshall McLuhan's concept of technological determinism and his contention that "We shape our tools and they in turn shape us" (Griffin, 2003, p. 344). Like any tool, technology is neither good nor bad; its value ultimately determined by use. For Millennials, and subsequent future generations, the tools of cybernetic technology will continue to influence human development, shaping everything from personal appearance to communication to identity.

In addition, Cyborg becomes a technological case study for young viewers. Not only watching the Teen Titan defeat the bad guys with his sonic cannon or superstrength but also seeing him struggle with his personal identity and humanity illustrates for children the advantages and potential drawbacks of our cyborg development. This is reinforced in several episodes, particularly "The Sum of His Parts," with the young boy who celebrates the prosthetic arm that makes him more like Cyborg. However, unlike the Titan who, at times, expresses resentment toward his transformation, the boy does not complain or curse his bad fortune; rather, he embraces technology as making him special. Again, to use Hegelian terms, he appears to have achieved balance or synthesis, becoming something of an idealistic role model for both Cyborg and the viewing audience, enthusiastically accepting the technological modifications of his own body and that of his titanium-plated hero. Additionally, the young boy supports Hutchby and Moran-Ellis' contention that the important question is not what impact technology has on children but how they relate to its varied forms. As the merger of body and machine continues to influence our concept of what it means to be human, understanding how children respond to this change will become even more crucial to not only their personal development but also the culture at large.

## Conclusion

From his first appearance in DC Comics in 1980 to his animated debut on the Cartoon Network in 2003, the teen superhero Cyborg has reflected the promise and peril of our technological age. As both human and machine Cyborg exemplifies the inherent tensions within this merger, struggling to maintain his identity within a body that is more metal than flesh and bone. Cyborg's internal conflict echoes that of another being created on the literary page centuries earlier, Frankenstein's creature. However, unlike Mary Shelley's daemon that distrusts humanity and seeks revenge on his creator, Cyborg learns to become relatively comfortable in his own cybernetic skin, using his abilities for the benefit of others. Moreover, Cyborg conveys a sense of optimism about our technological development, that while the body may become enhanced by or replaced with the artificial, we will always fight to keep those qualities that make us human. This character thus adds another page to the continually evolving Frankenstein myth, becoming another vision of *the* iconic image of our new millennium, the human/machine cyborg.

## References

Beecher, A. (2004). Only human (A. Soto, Director). In G. Murakami (Producer), *Teen titans*. Los Angeles: Warner Brothers.

Bloom, H. (1994). In M. Shelley, *Frankenstein*. New York: Penguin.

Britt, R.R. (2006). Wireless bionic arm would feel real. *LiveScience.com* Retrieved April 25, 2006, from http://www.livescience.com/humanbiology/060424_bionic_arm.html

Cadigan, G. (2005). *Titans companion*. Raleigh, NC: TwoMorrows Publishing.

Campbell, R., Martin, C.R., & Fabos, B. (2007). *Media and culture: An introduction to mass communication*. Boston, MA: Bedford/St. Martin's.

Clark, A. (2003). *Natural-born cyborgs: Minds, technologies, and the future of human intelligence*. New York: Oxford University Press.

Davis, R. (2006, September 14). Meet the $4 million woman. *USAToday.com* Retrieved September 15, 2007, from http://www.usatoday.com/tech/science/2006-09-13-bionic-arm_x.htm?POE=TECISVA

Dumit, J. & Davis-Floyd, R. (1998). Cyborg babies: Children of the Third Millennium. In R. Davis Floyd & Joseph Dumit (Eds.), *Cyborg babies: From techno-sex to techno-tots* (pp. 1–18). New York: Routledge.

Gray, C.H., Mentor, S., & Figueroa-Sarriera, H.J. (1995). Cyborgology: Constructing the knowledge of cybernetic organisms. In C.H. Gray (Ed.), *The cyborg handbook* (pp. 1–14). New York: Routledge.

Griffin, E. (2003). *A first look at communication theory*. 5th edition. New York: McGraw-Hill.

Halacy, D.S. (1965). *Cyborg: Evolution of the superman*. New York: Harper & Row.

Holland, S. (1995). Descartes goes to Hollywood: Mind, body and gender in contemporary cyborg cinema. In M. Featherstone, & R, Burrows (Eds.), *Cyberspace, cyberbodies, cyberpunk: Cultures of technological embodiment* (pp. 157–174). Thousand Oaks, CA: Sage Publications.

Howe, N. & Strauss, W. (2000). *Millennials rising: The next generation*. New York: Vintage Books.

Hughes, J. (2004). *Citizen cyborg: Why democratic societies must respond to the redesigned human of the future*. Jackson, TN: Westview Press.

Hutchby, I. & Moran-Ellis, J. (2001). *Children, technology and culture: The impacts of technologies in children's everyday lives*. New York: Routledge.

Lykke, N. & Braidotti, R. (1996). *Between monsters, goddesses, and cyborgs: Feminist confrontations with science, medicine, and cyberspace*. New Jersey: Zed Books.

O'Mahony, M. (2002). *Cyborg: The man-machine*. New York: Thames & Hudson.

Rushing, J.H. & Frentz, T.S. (1989). The Frankenstein myth in contemporary cinema. *Critical Studies in Mass Communication, 6*, 61–80.

Shelley, M. (1994). *Frankenstein*. New York: Penguin.

Slack, D. (2003). The sum of his parts (A. Soto, Director). In G. Murakami (Producer). *Teen titans*. Los Angeles: Warner Brothers.

Slack, D. (2005a). Go (M. Chang, Director). In G. Murakami (Producer). *Teen titans*. Los Angeles: Warner Brothers.

Slack, D. (2005b). Titans east: Part II (M. Chang, Director). In G. Murakami (Producer). *Teen titans*. Los Angeles: Warner Brothers.

Slusser, G. (1992). The Frankenstein barrier. In G. Slusser & T. Shippey (Eds.), *Fiction 2000: Cyberpunk and the future of narrative* (pp. 46–71). Athens: University of Georgia Press.

Thomas, D. (1995). Feedback and cybernetics: Reimaging the body in the age of the cyborg. In M. Featherstone and R. Burrows (Eds.), *Cyberspace/cyberbodies/cyberpunk* (pp. 21–43). Thousand Oaks, CA: Sage.

Wagner, J. & Lundeen, J. (1998). *Deep space and sacred time:* Star Trek *in the American mythos*. Westport, CT: Praeger.

Wilson, R.R. (1995). Cyber(body) parts: Prosthetic consciousness. In M. Featherstone & R. Burrows (Eds.), *Cyberspace/cyberbodies/cyberpunk* (pp. 239–259). Thousand Oaks, CA: Sage.

Wolfman, M. (2005). Titans East: Part I (A. Soto, Director). In G. Murakami (Producer). *Teen Titans*. Los Angeles: Warner Bros.

Wolfman, M. & Perez, G. (June, 1982). *Tales of the new teen titans, 1*(1). New York: DC Comics.

# New Frontiers: Cylons and Stargate; *Serenity* and the City

•CHAPTER FOUR•

# *Galactica*'s Gaze: Naturalistic Science Fiction and the 21st Century Frontier Myth

*Robert L. Strain, Jr.*[*]

The American frontier myth has been advanced by a variety of rhetorical forms, including novels, short stories, photography, dime novels, speeches, and films. Even before the birth of moving pictures, cutting-edge special effects created images that "moved" (emotionally if not actually) and depicted visual spectacles that gave shape, form, and meaning to the frontier myth. One of the influential methods of myth creation that utilized these effects, visual and otherwise, was nineteenth-century landscape painting. For more than fifty years this tradition helped spin an important ideological thread about the mythology surrounding America's move westward and the general philosophy of the American frontier. The Hudson River School of landscape painting is a prime example of the role of art in developing frontier mythology. Albert Boime argues that the Hudson River artists focused on "the assumed presence of the spectator, the horizontal format, the directional sense of gradual descent from a height to the valley and generally...from overgrown thicket to clearing or body of water to the trappings of civilization" (1991, p. 35). These paintings contained great sweeping vistas that positioned the viewer to see the American destiny of new land to conquer, tame, and occupy. This image of the spectator looking down over an awe-inspiring vista, what Boime calls the "magisterial gaze," is

[*] Dr. Robert L. Strain, Jr. (M.A., Baylor University; Ph.D., University of Kansas) is an Assistant Professor of Communication at Florida Memorial University.

"so pervasive in the art and literature of nineteenth-century America that we have to see it as an encoding of a solution to the American dilemma" (p. 10). It represents and furthers a desire to dominate the land and is an essential element in the mythology of the Western frontier.

This ideology of Western frontier mythology continues to appear in various art forms, including television. Jan Johnson-Smith (2005) suggests that the same magisterial gaze that was represented in the Hudson River School is an important component in made-for-television science fiction. Most made-for-TV space fiction in the United States and elsewhere looks forward, depicting space travel as an outward-bound adventure that materializes outer space in much the same way that the Western frontier was materialized in the narratives, politics, and art of the late nineteenth-century. For instance, the *Star Trek* series, in all of its renditions, is indicative of that tendency in science fiction television to look forward to an idealized future through exploration. My perspective, however, is that in our postmodern, post-9/11 world, a growing body of science fiction participates in a type of retelling of the traditional frontier myth. In particular my position is that while traditional renderings of made-for-TV science fiction through the late twentieth-century continued to utilize the magisterial gaze identified by Boime (and applied to science fiction by Johnson-Smith), a new way of articulating that myth is emerging.

Rhetorical scholars have long recognized the significance of myth in understanding human development. In her intriguing work on the frontier myth and the human psyche, Janice Hocker Rushing (1991) argues that, "Because the frontier myth is a powerful and value-laden embodiment of cultural identity, it is important for the rhetorical critic to trace its changes" (p. 244). This chapter considers those changes by exploring the cultural-rhetorical position of the television series *Battlestar Galactica* (*BSG*), treating the story not only as a reimagining of the short-lived series of the same name from the late 1970s but also as a signal of a shifting gaze in the American frontier myth. *Battlestar Galactica* uses forms of television production, direction, and visual aesthetics that construct a new point of view for the audience and creates a functionally different frontier myth from traditional expressions of science fiction on television. George M. Wilson's (1986) assessment of cinematic point of view in his book *Narration in Light* is helpful for my treatment of how *Battlestar*

*Galactica* breaks the traditional generic form of made-for-television science fiction and consequently redefines the form and function of the myth of the Western frontier.

Wilson (1986) develops a theory of point of view in narrative film that considers the audience's foreknowledge and narrative expectation to be a significant component of how they interpret the story. He writes,

> Viewers dispose their attention toward the 'focus' of the story's ultimate resolution. They are perceptually set to follow the evolution of those plot conflicts that are marked out to be the subject of an audience's most immediate and engaged regard. Indeed, it is the normal goal of narrative strategies to make this temptation effectively irresistible. However, there are narrative films...in which central aspects of their interest and significance bear only an *oblique* relationship to the forms of dramatic closure they employ. In these instances, the strong narrative orientation of viewer awareness and expectation tends to block the obliquely situated prospects entirely from view. That is, various factors that either appear on the screen or are implied but not shown in the film and that are peripheral to the strict development of the basic tale may be assigned a weight in the narration in such a way that the chief issues raised by the drama come to be modified, displaced, or otherwise reappraised. It is not too much to say that when this counternarrative weighting is apprehended, the whole gestalt of the film often seems to shift.... The problem for the viewer of such a film is to locate a 'centered position' from which the oblique strands of narrational strategy can come together in a configuration that reorganizes his or her perception and comprehension of the fictional events. (p. 10)

*Battlestar Galactica,* through its attempt at developing "naturalistic" science fiction, asks its viewers to reappraise the standard myth of the frontier as it is usually employed in made-for-TV science fiction. In other words, as *Battlestar Galactica* reorients the narrative elements of traditional made-for-TV science fiction, it shifts how frontier mythology is depicted in televised science fiction. The series' "inward" focus on the lives of its characters and the use of "realistic" visual imagery and narrative content shift the magisterial focus of the spectator, from outward to inward, and facilitate the development of a different type of frontier myth for the twenty-first century.

My application of Wilson's basic perspective on point of view follows the general direction of Ben-Tov's (1995) rejection of the commonly held notion that science fiction as a genre must take into consideration the mythology of the American frontier. Ben-Tov (1995) writes,

> It is not enough to say that science fiction is like the western or is unlike it or is influenced by it. Science fiction combines frontier mythology with myths about collective technological power. It does so in an effort to define an American (white) identity, using the frontier myth to connect two aspects of American experience: the natural wilderness and the technological system. (p. 9)

Rather than demonstrate how *Battlestar Galactica* is like or unlike a Western, I use my interpretation of Ben-Tov's perspective to test *Battlestar Galactica*'s expression of frontier mythology. Wilderness (outer space/inner space) and technology create a narrative that reflects the evolving relationship of Americans to the frontier. Additionally, I explore how the reality TV phenomenon has influenced *Battlestar Galactica*'s presentation of naturalistic science fiction and explain what that means for the evolution of frontier mythology in the twenty-first century.

## Naturalistic Science Fiction and Reality Television

Since its original airdate on the Sci-Fi Channel in December 2003, the "reimagined" *Battlestar Galactica* has developed a growing, enthusiastic following. The original pilot has turned into a popular series that some argue is changing the genre of science fiction. Without a doubt the series has raised a great deal of popular and critical enthusiasm. *The New York Times Magazine* states that *Battlestar Galactica* is "one of the most original and provocative programs on television" (Hodgman, 2005, p. 34). One important reason for this is the series' move away from what the series producer David Eick refers to as the "trappings of the [sci-fi] genre" to tell its story. The central elements of traditional science fiction storytelling are largely ignored and replaced by a "naturalistic" approach. "We take as a given the idea that the traditional space opera, with its stock characters, techno-double-talk, bumpy-headed aliens, thespian histrionics and empty heroics has run its course, and a new approach is required" (Hodgman, 2005, p. 36), says the show's writer-producer Ron Moore. Consequently, *naturalistic science fiction* uses docu-style camera work, the explicit presence of the camera in the film, realistic situations, and nonsanitized interactions among characters. The stories told in the series function through the "reality" of the visual images on the screen, the believability of obviously fictional situations, and the performances of their characters.[1] The conventions of televised science fiction that have dominated the past forty years of broadcast television are replaced with an obviously different form and content.

Through its nontraditional use of visual forms and narrative content, *Battlestar Galactica* frames its story in a significantly different fashion from most traditional space fiction. The "magisterial gaze," that Johnson-Smith (2005) argues is a part of science fiction television is clearly absent from *Galactica*. The sensibility and point of view of the series are not on the outward, forward-looking arc of traditional American science fiction. There are no typical frontier hero "types," who are a staid element of so much made-for-TV space fiction. Rather, the "gaze" of *Galactica* emphasizes an inward-looking, measured point of view in which characters who appear to be heroes actually have a more complicated place in the story. The emotional, psychological, and relational situations of characters take a much more primary place in the development of the narrative. Moore and Eick (2005) continually stress their desire to make the series a "drama" first and a science fiction show second. Much less emphasis is placed on concerns of seeking out "new life and new civilizations," which was such an important feature of the *Star Trek* franchise, for example. This inwardly focused gaze features the actions of the characters and the performances of the actors. For instance, director Michael Rymer mentions the controversy surrounding the way that actor Jamie Bamber portrays Apollo in the miniseries. Apollo's character is apparently a hero. However, Bamber's portrayal is nuanced in such a way that the character is not particularly likable and has a strained relationship with his father—breaking traditional expectations of the character for those who knew the original series as well as the general expectations audience members may have for a hero character. This performance (and others in the series to be examined later) adds a sense of "reality" to the story. Eick (Moore & Eick, 2005) argues that the performances of the actors along with all the other elements of production contribute to the notion of reality and is "naturalistic, it's not theatrical, it's not done in a vein that calls your attention to it. It's the kind of conversation that you might hear every day." This allows the audience to establish familiarity with characters in a more fundamental way than they might with stock characters in traditional televised science fiction series.

It is my position that the shift in emphasis of form and content displayed by *Battlestar Galactica* has been influenced by the rise of reality television, a medium that has changed how we watch televised fictional entertainment today. Recent scholarship on reality television reveals a great deal of ferment in the emerging definition of this form of entertainment. [2] While it is well beyond the scope of this chapter to define the entire genre of reality TV, it is important to identify how I apply reality TV to a study of science fiction and American frontier mythology.

The relationship of myth to reality has long been a part of developing thinking on myth. From Malinowski's (1954) view of myths as a "reality lived," to J.O. Robertson's (1980) juxtaposition of reality and myth in *American Myth, American Reality*, to more recent treatments of this relationship such as Jon Wagner and Jan Lundeen's (1998) *Deep Space and Sacred Time:* Star Trek *in the American Mythos*, myth and reality have historically been intertwined. The work on landscape painting mentioned above also reflects this relationship and provides a significant example of the way that artistic form can influence social mythologies. Boime (1991) further illuminates his concept of the magisterial gaze:

> It presupposes the spectator as sightseer on the ledge or crest subjugating the boundless *reality* to a disciplined scrutiny and simultaneously taking a reading from this orientation that is profoundly personal and ideological at the same time. The panoramic prospect becomes a metonymic image—that is, it embodies, like a microcosm, the social and political character of the land—of the desire for dominance. (pp. 20–21, italics mine)

Landscape painting of the nineteenth century contained formal elements such as audience point of view and visual style whose rhetorical force clearly affected the development and direction of frontier mythology. The ideology contained in many of those painted images (and the viewer's participation in that gaze) illuminated a frontier perspective that influenced how Americans conceptualized the West. Reality television is having a similar influence on the *form* of television production and aesthetics that may be reorienting our gaze toward the myth of the American frontier. Current stories of the West are "filtered" through the lens/gaze of reality television, spinning an emergent view of the West. Today's television viewers have a heightened *expectation of the real* resulting from the saturation of reality TV. Such an expectation demands that the visual and narrative aesthetic of fictional television appear authentic. In other words, even if a program is clearly fictional it should at least have the appearance of being "real." Both the technological and discursive characteristics of these programs are expected to ring true for the spectator. The technical aspects of filmmaking require that art and technology imitate life as closely as possible. Narrative storylines are expected to be within the bounds of possibility and the actions of characters must be written and performed with a sense that they could indeed be those of real people.

Though the treatment of what is "reality" differs between the various reality TV formats, there is an important similarity: the audience is aware of the presence of the camera. Sometimes the subjects are aware that the camera is present; sometimes they are unaware. However, the *audience* is always made aware (directly or indirectly) that the camera is present. Today's television audiences, then, have an enhanced sense of point of view regarding filmed images. The awareness of that gaze in these types of programs creates in the audience an *expectation of the real*, a sense that some element of what we are viewing is "true to life," at least to some degree—even in programs known to be fiction. Images that are "true to life" are not necessarily empirically true. Rather they are functionally true to the viewer as they echo some *sense* of the real held by the viewer. Such sensitivity to "reality" stems from the growing familiarity of television audiences with documentary realism constituted by two elements: observational realism and expositional realism. These concepts "ask the audience to register the techniques used to observe real life (for example, handheld cameras), or the way in which an argument is presented to us (for example, interpretation of evidence)" (Hill, 2005, p. 19). While the truth claims of the "realness" of reality television have been hotly debated over the years, no one doubts that the camera sees something. It is those viewed images that audience members accept as real on the basis of the expectation arising from familiarity with the general grammar and rhetoric of reality television.

## *Battlestar Galactica* and the Expectation of the Real

*Battlestar Galactica* creates an expectation of the real through its visual aesthetic and performative characteristics. The first place in which the series displays its uniqueness is in the visual imagery. The program crafts a visual grammar that breaks the prevailing convention in filmmaking that suggests audiences should not feel the presence of the camera. In his DVD commentary to the pilot for the series, Moore (Moore & Eick, 2005) discusses his ideas about making *Battlestar Galactica* a more "realistic" form of fictionalized television. He addresses the presence of the camera:

> There's this theory in film about never wanting to feel the camera, you want to be in a scene and you want the audience to sort of feel like they are there on some level, you never want to be aware of the camera, and this is sort of taking the opposite approach. You are creating something that is fundamentally unreal, and the only way, in my opinion, you convince an audience it is real is by giving them a sort of underlying sense that there is a camera involved, there is somebody who had to actually go out and shoot this and by providing them with that idea, they then buy into the visual aspect of what they are watching in a more fundamental way.

From a technical, directorial point of view, the producers make a conscious attempt to imbue the story with a sense of reality. Placing the audience in closer sensory proximity to the on-screen action (a characteristic of the reality TV aesthetic that has developed over the past fifteen years) decreases the distance audiences may feel when watching traditional science fiction. For *Battlestar Galactica,* keeping a visual sense of reality, the feeling that the characters are actual people aboard an actual spaceship doing actual things means that the audience *must feel* the presence of the camera. The visual look of the series is substantively different from most made-for-TV space fiction before it. For instance, in one of the earliest shots in the miniseries, a piece of a space station that has been destroyed by a missile seems to hit the camera, knocking it from its fixed position and giving the audience the sensation that their point of view has actually been hit by the debris. In battle sequences throughout the series, especially the action of spacecraft in space, there is frequent use of a "zoom" effect when the camera views the action from a long distance and then closes in quickly on the subject.[3] This type of camera work creates a sense of realism for the viewer as it follows a tradition of combat photography and its evolution into fictionalized combat scenes.

## Reality and Cognitive Estrangement in *Battlestar Galactica*

The distinctive visual form, the thematic content, and the performance of the actors in *Battlestar Galactica* resonate with audiences because of how they display what Darko Suvin (Gunn & Candelaria, 2005) calls cognitive estrangement. Cognitive estrangement functions by arranging elements of the familiar (and apparently "real") with elements that are re-fictionalized but fundamentally unreal. Suvin argues that science fiction is "the literature of cognitive estrangement" (Gunn & Candelaria, 2005, p. 24) existing along a continuum of "estrangement" and "cognition"—essentially, from the familiar to the unfamiliar. The term "estrangement" is borrowed from Formalist literary criticism, which developed an approach to estrangement that emphasized the power of presenting material that was simultaneously familiar and unfamiliar. Suvin quotes Bertolt Brecht to make his point when he says, "A representation which estranges is one which allows us to recognize its subject, but at the same time makes it seem unfamiliar" (Gunn & Candelaria, 2005, p. 25). Cognition focuses on that which is familiar. It "implies not only a reflecting *of* but also *on* reality" (p. 29). Balancing the unfamiliar with what is knowable and familiar is central to cognitive estrangement and the larger goal of speculative fiction. I would like to consider now the characteristics of both cognition and estrangement and explore how they function in *BSG*.

Estrangement and cognition carry the narrative and ideological elements in social mythologies. Their interaction gives force to science fiction as a literary genre. I suggest that science fiction in general and *BSG* in particular function in this way and are developing an updated myth of the American frontier. I am not the first to suggest that cognition and estrangement work together in science fiction.[4] However, I contend that the influence of reality television, at least in the case of *Battlestar Galactica*, demands that we consider new lines of analysis for the frontier myth. To the extent that today's television audiences have a heightened expectation of the real, the emphasis that *Battlestar Galactica* places on "reality" means that it may be possible to look at its newly rendered science fiction narrative and evaluate how it carries various social myths.

*Battlestar Galactica* reorganizes the myth of the American frontier. Traditionally, that myth has emphasized the centrality of larger-than-life heroes "struggling against natural and supernatural forces to create an economically progressive and dynamic civilization in an unknown wilderness" (Dorsey, 1996, p. 46). Nineteenth-century landscape painting certainly featured this ideology as a central part of its imagery. Moreover, the traditional science fiction films and television series from the second half of the twentieth century owe a significant debt to that expression of the mythology of the Western frontier.

Richard Slotkin (1992) argues that science fiction film and television work like fairy tales to "allegorize the condition and etiology of the present world, but they purchase imaginative freedom—the power to imagine the most magical or utopian possibilities—by keeping real historical referents at a distance" (p. 636). Conversely, *Battlestar Galactica* actually makes clear references to historical events and puts a realistic spin on its treatment of these events.[5] In addition, the series recasts the treatment of wilderness and technology in the frontier myth in a manner that responds to the national audience's expectation of the real and follows a growing trend in the genre to address the important everyday realities of life in a frontier community (and focuses on the "inner space" of heroes). For example, in his work on science fiction literature's treatment of homesteading, Carl Abbott (2005) concludes more recent trends in the literature move away from the traditional model that frames homesteading novels such as family sagas that fit into the master mythology of the West, which emphasizes taming the land. Abbott identifies several science fiction novelists whose stories treat homesteading with a complexity that elucidates its role in the larger process of social change in America. These novels move away from the allegorical treatment of traditional issues to a much more sophisticated and realistic treatment of sociocultural history. This trend in literary science fiction, Abbott suggests, displays the genre's ability to help explore larger and deeper questions and issues in American frontier mythology. He states, "sf [science fiction] homesteading stories are most challenging when they step beyond the frames of adventure tale and family saga to place homesteading within larger narratives of economic development and cultural transfer" (p. 260). Abbott couches an important aspect of his argument about homesteading science fiction literature in the influence of New Western History and its way of "retelling" the Western frontier myth. The important development of New Western History during the 1990s may also help us understand how a new myth of the Western frontier might evolve in made-for-TV science fiction.

Patricia Nelson Limerick, Clyde A. Milner II, and Charles E. Rankin (1991) summarize New Western History as offering a "more balanced view of the western past." They state,

> It [New Western History] includes failure as well as success; defeat as well as victory; sympathy, grace, villainy; and despair as well as danger, courage, and heroism; women as well as men; varied ethnic groups and their differing perspectives as well as white Anglo-Saxon Protestants; an environment that is limiting; interactive, and sometimes ruined as well as mastered and made

> to bloom; a parochial economy alternately fueled and abandoned by an interlocking national and world order; and finally, a regional identity as well as a frontier ethic. (p. xi)

New Western History, then, is a kind of bona fide revisionism steeped in the reality of the "true" American experience of westward expansion.[6] This focus on the tough realities of Western American history coupled with the growing place of television in general as a means of storytelling and the appeal of some of reality television's technical conventions in particular would indicate at the very least that audiences today might be open to new ways of listening to narratives, especially those that might draw on "re-envisioned" elements of our mythic past. My position is that *Battlestar Galactica*, with its energized *expectation of the real* and its narrative use of elements of the Western frontier myth such as technology and the wilderness articulates a "revised" myth of the Western frontier.

## Technology, Wilderness, and *Innerspace*

In her article, "Frontierism and the Materialization of the Psyche: The Rhetoric of *Innerspace*," Rushing (1991) explores the question of how a frontier hero might respond to explorations of internal, psychological, spiritual, and moral elements of his own psyche. She analyzes the hero in the 1987 film *Innerspace* in order to test a hypothesis about "how...the frontier hero [will] act when exploring the scene of inner space" (p. 245). She concludes, as she has in some of her previous work on myth in science fiction films, that any impulse toward introspection into the internal elements of the psyche that might evaluate deeper questions of spirituality, morality, or contemplation is co-opted by the frontier myth's emphasis on action rather than introspection. Rushing (1991) argues that,

> The continued dominance of extraverted [sic] sensation in the frontier myth perpetuates the direction that Western culture has been taking at least since the Renaissance—a movement away from the 'other world' of the spirit and toward the 'real world' of the physical.... The outer, material world has long been recognized as the only 'reality' sufficiently valid to garner serious attention. Furthermore...Western humanity has increasingly transferred its reverence from the supernatural to itself, if not as the creator, then at least as the subjugator of the material world. (pp. 253–254)

Along the continuum of inner space—outer space, outer space and the centrality of action in that outer space trump any thorough introspective exploration of the psyche of the hero and uphold the traditional American frontier myth. This section of the chapter explains how *BSG* overturns this characteristic that Rushing reveals in her analysis. *Battlestar Galactica*'s narrative use of technology and the notion of wilderness allows us to reconsider how inner space is treated in made-for-TV science fiction.

## Technology

The opening minutes of *BSG: The Miniseries* frames its treatment of technology and space (wilderness) and informs the observer that it will be a different treatment from traditional science fiction's typical emphasis on "the conflict between pastoralism, the ideal of America as a vast garden, and the rhetoric of the technological sublime, which justifies industrialism and technological expansion" (Ben-Tov, 1995, p. 9). *Battlestar Galactica* presents technology in a fashion that strikes a chord of familiarity in viewers. Most recent examples of technology in science fiction television serve the traditional myth of the frontier and move toward conquering the "vista." This is the first place where we can begin to see how *Battlestar Galactica* starts to remake the frontier myth. Technology is not featured as the primary means of liberation and dominance of "space" (material and otherwise). Rather, technology has a familiarity that allows the audience to focus more on the other elements of the story without calling attention to itself. In this fashion the series begins to bring a different gaze to bear on the frontier and its mythology. The narrative's reoriented relationship to technology is visible from the earliest scenes of the series.

The story opens with title cards placed over the beginning action of the film. These are explanations about the relationship between humans and the "race" of robots called Cylons created to "make life easier for the twelve colonies" of humans. Such a revelation shows an immediate role for technology in the narrative arc of the miniseries itself and the series as a whole. The sequence establishes that the Cylons rebelled against their human creators and after a long conflict fled to another world. The story begins with the Cylons returning to apparently annihilate the human race. This sort of inability to control technology is not a unique theme in science fiction; however, as the miniseries develops we begin to see that technology is framed very differently. For example, *Battlestar Galactica* displays for its viewers familiar-looking technology. First, items such as telephones, paper, and even dot-matrix printers are similar to technology on twentieth-century Earth. Guns shoot bullets. Calls are made with telephones that have cords. Food is grown, prepared, and eaten from plates. Computers are networked. These aspects of familiar technologies are prominent throughout episodes in the series. In fact the series' production team purposefully establishes this familiarity in order to create a greater sense of reality and believability. Such familiarity functions to add a layer of reality to the series. The viewer need not focus attention on trying to understand the protocol of unfamiliar technology. This strikes an important balance between estrangement and familiarization in terms of how the audience relates to technology. Moreover, it allows a myth's ability to address social problems to attain more rhetorical force in a narrative.

Characters in the series also help frame the relationship to technology. Early in the pilot episode we learn that the commander of the *Galactica*, Adama, will not allow the computers on the vessel to be networked (his decision means the ship is immune to the jamming technology that the Cylons use to shut down anything that uses networked computers and that facilitates their destruction of the twelve colonies). One of the miniseries' most powerful examples of the relationship between technology and space is the speech that Adama gives at the decommissioning ceremony for his ship. Just before he arrives at the ceremony the audience is introduced to Adama's strained relationship with his son, Apollo, who is also a pilot in the military (the Colonial Fleet). As he begins the speech, Adama is clearly preoccupied with the relational problems he is having with his son. Before finishing the first sentence he deviates from his prepared text and gives an oration about the "worthiness" of humankind to exist. He says,

> Adama: The Cylon war is long over. Yet we must not forget the reasons why so many sacrificed so much in the cause of freedom. The cost of wearing the uniform can be high but sometimes it's too high. You know when we fought the Cylons we did it to save ourselves from extinction, but we never answered the question "why?". Why are we as a people worth saving? We still commit murder because of greed, spite, jealousy. We still visit all of our sins upon our children. We refuse to accept the responsibility for anything that we've done. Like we did with the Cylons. We decided to play God. Create life. When that life turned against us we comforted ourselves in the knowledge that it really wasn't our fault, not

> really. You cannot play God and then wash your hands of the things you have created. Sooner or later there comes a time when you can't hide from the things you've done anymore. (Moore & Eick, 2005)

When Adama shifts direction/focus in the speech in the second line and states that the price of wearing the uniform can sometimes be "too high," he is unabashedly addressing the internal questions that Rushing's (1991) analysis concludes are co-opted by the extroverted actions of the hero. Adama's themes clearly display, for a moment at least, a consideration of introspective contemplation of "inner space." My view here is that this rhetorical artifact displays what may be a type of beginning of the end of the "dominance of extraverted [sic] sensation in the frontier myth" (Rushing, 1991, p. 253). The ideas in the speech signal how the series actually turns its focus to the psyche of "heroes" in what may possibly be a "new" frontier myth. For example, there is both a warning and a call for self-reflection when Adama ends the speech by saying "sooner or later there comes a time when you can't hide from the things you've done anymore." This comment disregards the triumph of humanity over technology and refers to both the mistakes of his personal life as well as humanity's hubris in creating the Cylons. Adama's reference to creating (artificial) life is a telling reference to humanity's dependence on technology. In addition, his standing order not to network his vessel's computers is a familiar fear that many people currently feel. Adama equates this use of technology with a sin that humanity must eventually address.

The speech is an extraordinary departure from the themes of most televised American science fiction. The audience's gaze is shifted from the forward-looking focus on the sublime that dominates most televised science fiction and toward introspection about limits, weaknesses, and failures of humanity. Adama even questions the worthiness of humanity to escape extinction. Shortly after the decommissioning ceremony, Adama and his crew learn that, after forty years of peace, humanity faces extinction again at the hands of the Cylons. From this moment, the miniseries radically shifts gears and becomes what is apparently a traditional sci-fi space battle film. During the first major battle scene, there is another reference to the limitations/problems with technology. The Cylons have infiltrated the defense system of the twelve colonies and can now turn off any Colonial ship that has networked computer systems. Only ships without networked systems or older spacecraft are immune, making Adama's decision look wise rather than old-fashioned.

## Wilderness

One of the essential elements of American landscape painting of the nineteenth-century is the unique celebration of the wilderness. Television series such as *Star Trek* displayed a similar kind of reverence for outer space. For *Battlestar Galactica*, outer space is treated very much as a brutal wilderness. Exploration of that wilderness is done solely for the purpose of survival, neither to celebrate the sublime nor to gain scientific knowledge. In another speech to his crew at the end of the miniseries, Commander Adama once again shifts the gaze of the narrative to another important aspect of American frontier mythology: the wilderness (outer space). In a speech at a funeral ceremony mourning their comrades killed in battle, Adama essentially completes a prayer begun by a priest at the ceremony. After sensing his crew's low morale, he asks them, pointing to the coffins, "Are they the lucky ones?...Maybe it would have been better for us to have died quickly, back on the colonies with our families. Instead of us dying out here, slowly, in the emptiness of dark space." Adama's words here point to the sobering reality of their situation: outer space is not a place that creates the kind of sublime that demands a magisterial gaze. It is empty and dark. It is a means to an end. It is a trial to endure. In fact, there are several instances during the battle scenes when the camera shows crewmembers and fighter pilots dying in space, without protection of their spacecraft. They die, exposed in their space suits, in the vacuum of space.

Adama ends this speech by revealing a "secret" involving a mythical planet called Earth, where a thirteenth tribe of colonists landed. His mention of the "legend" of Earth, we later discover, is a ruse that he uses to instill a sense of hope in the demoralized crew and fleet. This attempt to introduce a sense of hope in the fleet is significant because it begins to develop the series' method of balancing the relationship of the crew to their plight in the "wilderness" of outer space with how they will deal with each other in what I call "inner space." The miniseries illustrates this balancing immediately after Adama finishes this speech. The audience sees nine intercut scenes that depict the crew returning to their life in the community of the spacecraft where they live. Not all of the relationships are without difficulty. In fact, the first scene we see after Adama's rousing funeral speech is one showing the continued discord between the executive officer of the *Galactica*, Colonel Tigh, and the ship's best fighter pilot, Lt. Kara Thrace. The other scenes all depict an attempt by crewmembers to return to some sense of normalcy after the trauma of experiencing the end of life as they know it.

## Wilderness and *Innerspace*

*Battlestar Galactica*'s treatment of the idea of "wilderness" demonstrates how reality television and the expectation of the real influence the presentation of made-for-television speculative fiction. One important feature of a great deal of reality television is to present "frontstage" and "backstage" dimensions of human experience. For example, many reality television programs feature some version of a confessional time—a moment that subjects use to "confess" to the viewer or other participants their "true" feelings and thoughts. The viewer develops a sense that often the backstage actions and thoughts of characters are more "real" that what is performed frontstage. This characteristic of reality TV demonstrates a willingness to feature the introspective aspects of individuals and the foregrounding of backstage behavior as more representative of the "real." So even within fictional television things seem more real when the audience is allowed "behind the scenes."

The first three episodes of season one of *Battlestar Galactica* explore the fleet's relationship with the wilderness and its impact on the realities of life within community. In each of these episodes, the audience experiences the juxtaposition of the narrative of the characters' inner life with their public life. In addition, traditional notions of the behavior of heroes in the Western frontier myth are overturned. In one way or another, the episodes "33," "Water," and "Bastille Day" all explore the impact that life in the wilderness of outer space has on the private and public life of the fleet. These episodes create the tenor for the entire series.

In "33" the fleet is under attack by the Cylon enemy every thirty-three minutes. This episode guides the spectator-audience's gaze toward wilderness that certainly resembles what Adama called "the emptiness of dark space." The first image we see is of the character Gaius Baltar dreaming—he is clearly fatigued. Baltar is considered a genius, one of the greatest scientific minds on Caprica; we are introduced to him as the one human who has clearly colluded with the Cylons before the original attack on Caprica. The next several scenes show the crew fighting off exhaustion, attempting to remain awake and alert. We soon learn that, in fact, they have been continually "jumping" away from their Cylon pursuers every thirty-three minutes for the past five sleepless days. This aspect of the episode portrays a strong sense of isolation that is the reality of outer space. The practical matters of this wilderness are made explicit in the next episode, "Water," when the fleet loses virtually all of its drinkable water through an act of sabotage and must begin searching for water in the wilderness. After locating water on a previously uncharted moon, the civilian and military commanders decide to use prisoners to supply the hard labor to extract the water. This storyline is examined in the episode titled "Bastille Day." Men detained on a prison ship are asked to "volunteer" to do the work on the surface of the moon. They decline, a prisoner rebellion ensues, and the resulting conflict raises significant questions concerning life in community that the surviving members of the human race must confront. The background for these introspective questions of a personal and public nature is the experience of the fleet in the wilderness of outer space. It is clear in the narrative frame of this series that humanity's experience in outer space does indeed have a impact on the inner life of individuals; *Battlestar Galactica*'s treatment of the narrative turns the audience's attention to these introspective elements in ways that traditional science fiction storytelling on television has historically ignored.

As the very practical matter of survival is the central focus of the series and is consistently counterbalanced by an ongoing treatment of the inner lives of the main characters of the drama, the relational dynamics of fleet life and the introspective experiences of several characters are prominently featured in the narrative. Each of the first three episodes of season one positions the audience to reflect on the sense of physical space where the action is located as well as the everyday experiences and emotional changes of these characters.

For example, when "33" opens with Baltar dreaming, he is conversing with a Cylon character, wondering whether she is a figment of his subconscious or a microchip implanted in his brain. The first words of dialogue in the episode are spoken by this Cylon character, a woman known as Number Six, to Baltar. She says, "God has a plan for you, Gaius. He has a plan for everything and everyone." Such an apparent concern for the lives of "everyone" is clearly made an important characteristic of *Battlestar Galactica*. An important element of the audience's relationship with Gaius is our understanding (or misunderstanding?) of his emotional, psychological, and moral status. Number Six is a continuous presence for Gaius's character; without

Number Six we do not have full view of his true character. To that extent, the audience never really knows whether she is "real" or merely a figment of his imagination. Either way, she serves as our means of viewing the inner dialogue and introspection of this character. In another crucial episode, called "Hand of God," we see Baltar tell the Colonial fleet's military commanders where to bomb a fuel depot in order to win a crucial battle against the Cylons. Though he asks Number Six for help (divine intervention) in choosing the right spot to attack, Baltar believes that he simply "picked that spot at random" and made a wild guess as to the targeting information. God did not tell him the location. To this, Number Six replies, "He doesn't always speak in words, Gaius." After the decisive victory, Number Six uses scriptures to convince Baltar that God wanted him to destroy the Cylon base and guided his finger to the target in order to fulfill that purpose and help bring on an even larger, still unrevealed destiny. By the end of the episode Baltar utters in quickly dissipating disbelief, "I am an instrument of God."

The episode "Hand of God" also begins a focus on the psychological/emotional transformation of another character, Laura Roslin. Roslin is introduced early in the miniseries pilot as the Secretary of Education of the Colonial government, and we meet her as she learns she has terminal breast cancer. After the initial Cylon attack on the colonies, she discovers that she is the only surviving member of the presidential cabinet. Though she is forty-third in line for the presidency, she is sworn in within hours of the Cylon attack. Through the progression of the season Roslin must exercise leadership under extreme personal duress. We see her confide her medical condition first to Apollo, later to a doctor, and finally to a priest. All the while Roslin must look within herself and weigh the morality of many of her decisions. For example, in the season's first episode, "33," she agrees to have a civilian spaceship that appears to be threatening the fleet destroyed. She wrestles with guilt associated with the decision and battles with the legitimacy of her presidency and her personal and public authority as leader. But it is in "Hand of God" that Roslin (and the audience) starts clearly to change psychologically. When the episode begins she is conducting a press conference during which she sees twelve snakes at her podium. She later tells a priest about her hallucination/vision of snakes and earlier dreams she has been having. The priest informs Roslin that the "sacred scrolls" predict the exile and rebirth of the human race. Quoting the scrolls, the priest says, "And the Lords anointed a leader to guide the caravan out of the heavens to their new homeland. And unto the leader they gave a vision of serpents numbering two and ten as a sign of things to come." This revelation to Roslin begins a personal transformation in the character as the story continues. In "Kobol's Last Gleaming, Part I," Roslin has another vision that the same priest confirms is based in the scriptures. After pausing to come to an apparent self-realization, Roslin replies, "It's real. The scriptures, the myths, the prophecies. They're all real." It is at this moment that Roslin's inner dialogue is also clear to the audience. She has realized her place in this story of rebirth for the human race. The narrative has taken time over several episodes to bring the audience along the way in Roslin's internal journey. The point of view given to the audience allows this time in Roslin's "inner space" to help us see the different emphasis of this frontier myth.

*Battlestar Galactica* purposely turns the audience's gaze to the introspective dimensions of the hero's psyche. Through characters such as Roslin and Baltar, we discover a depth to the personal lives of heroes that, in many ways, reflects the complexity of the frontier myth itself. Simplistic stories of good versus evil are replaced with narratives with more intricacies. When this occurs the story provides a more penetrating evaluation of the relationship of culture to the wilderness and technology. This challenges Rushing's (1991) conclusion that there is a split in the cultural psyche in which "extraverted [sic] sensate thinking" dominates "introverted intuitive feeling" (p. 254). Though characters such as Adama and Laura Roslin still make decisions that have a bold action-oriented result, *Battlestar Galactica* presents the intuitive feelings of the narrative's heroes in such a fashion that their place in collective and individual decision making (and the overall flow of action in the narrative) is clearly visible and unmistakably important. Reality television has popularized the docu-style filmmaking that is clearly a central part of the way *Battlestar Galactica* tells its story. This more internalized form of storytelling in science fiction television indicates that mythical stories of journeys can also make powerful statements about societies. Instead of constantly looking forward in order to dominate the space on the horizon, this new framework for telling the frontier myth looks internally at the realities of everyday personal experiences.

## New Frontiers in the Western Myth

*Battlestar Galactica* has important implications for understanding how televised science fiction can reenvision American frontier mythology and how trends in narrative television more generally inform us about the way social myths are understood. The master narrative of the series centers on a topic that is not that unusual in science fiction stories: a group of travelers find themselves lost in the "wilderness" of outer space. However, as demonstrated in the preceding analysis, specific elements of that narrative are turned on their heads. Because *Battlestar Galactica* turns its gaze on the brutality of the wilderness and the limitations/dangers of technology, it is different from many traditional expressions of space fiction on television. Even more importantly, the mythic journey through the wilderness, and the heroic behavior of characters on that journey, are imbued with a complexity that transcends the traditional tenets of Western frontier mythology. Though there remain plenty of action, battle scenes, a formidable foe, and even a reinvigorated sense of hope and perseverance, there is no sense of wonder, no sublime, in this narrative. Heroes are not heroic in the traditional sense. The story is character-driven rather than guided by an ideology of "The West" or space as a New Frontier. As a result, characters who might be portrayed as heroes in typical made-for-television science fiction series are often less than awe-inspiring. Baltar's god complex, the health of Laura Roslin that unexpectedly leads to her newfound faith in religion, and the cynicism of Adama are all examples of *Battlestar Galactica*'s willingness to focus the audience's point of view on elements that, on the surface, have little to do with the resolution of the main narrative of the journey to find Earth. The main revelation here is that the audience's gaze has been turned to the actual "inner frontier" that centers on the internal motivation of characters who are very much (too much?) like regular people instead of heroes. The inward gaze of *Battlestar Galactica*, then, highlights crew life/relationships, social dilemmas, ethical questions, and the psychological well-being of characters at the expense of the magisterial gaze projected in traditional science fiction. Instead, the biting reality of another point of view takes precedence. The result is a reoriented interpretive position that allows the broad framework of the traditional lost in space science fiction narrative to function differently, a new mythological perspective for a new millennium.

This twenty-first-century frontier myth focuses on the following:

1. an internalized gaze that turns the audience's point of view onto specific elements of the tensions between individuality and community
2. a "humanized" hero
3. an intensified relationship between the familiar and the unfamiliar.

Essentially, *Battlestar Galactica* moves to purge the "frontier trappings" of traditional American West mythology from televised science fiction. This analysis suggests that the form and content of traditional frontier mythology may change as the audience's ability to understand and participate in their own sociocultural history changes. Reality television and New Western History provide examples of how the perceptual lens that audiences use to understand cultural narratives changes over time. As the audience gets more familiar with television as a form of storytelling and with the content of the stories the medium portrays, it develops a dynamic relationship with sociocultural narratives and myths. Reality television and New Western History have combined to create a powerful sense of expectation of the real in viewing certain televised narratives. The influence of reality television is significant here because it provides an expectation of transparency to stories. Though the stories are obviously fictional, audiences are treated to production elements that are characteristic of the dominant reality television genre, and so we buy into the "realness" of the story. *Battlestar Galactica*'s place as an example of naturalistic science fiction has essentially allowed the televised science fiction genre to catch up to some of the tenets of New Western History that help retell the myth of the Western frontier.

## Notes

1 For a full discussion of naturalistic science fiction by the series producers, see the miniseries DVD with commentary.

2 Examples of reality TV include (but are not limited to) reality talent formats, lifestyle formats, reality game formats, infotainment formats, surveillance formats, fly-on-the-wall formats, docu-soap formats, reality life experiment formats, and reality clip-show formats.

3 Thanks to David Whitt for help with identifying this visual aspect of *Battlestar Galactica.*

4 Darko Suvin was the first to theorize the application to science fiction.

5 The issues and historical references that *Battlestar Galactica* makes clear allusions to are terrorism, genocide, the limits of technology, and religious fundamentalism.

6 For an excellent summary of New Western History, see Worster (1991).

# References

Abbott, C. (2005). Homesteading on the extraterrestrial frontier. *Science Fiction Studies, 32,* 240–264.

Ben-Tov, S. (1995). *The artificial paradise.* Ann Arbor: University of Michigan Press.

Boime, A. (1991). *The magisterial gaze.* Washington, DC: Smithsonian Institution Press.

Dorsey, L. (1996). The myth of war and peace in presidential discourse: John Kennedy's "new frontier" myth and the Peace Corps. *The Southern Communication Journal, 62,* 42–55.

Gunn, J. & Candelaria, M. (2005). *Speculations on speculation: Theories of science fiction.* Lanham, MD: Scarecrow Press.

Hill, A. (2005). *Reality TV: Audiences and popular factual television.* London: Routledge.

Hodgman, J. (2005). Ron Moore's deep space journey. *The New York Times Magazine,* July 17, 32–37.

Johnson-Smith, J. (2005). *American science fiction TV:* Star Trek, Stargate *and beyond.* Middletown: Wesleyan University Press.

Limerick, P., Milner, C., & Rankin, C. (1991). *Trails: Toward a new western history.* Lawrence: University of Kansas Press.

Malinowski, B. (1954). *Magic, science and religion.* Garden City, NY: Doubleday.

Moore, R. & Eick, D. (Executive Producers). (2005). *Battlestar Galactica* (DVD—Season One/Disc One). Universal City, CA: Universal Studios.

Robertson, J. (1980). *American myth, American reality.* New York: Hill and Wang.

Rushing, J. (1991). Frontierism and the materialization of the psyche: The rhetoric of *Innerspace. Southern Communication Journal, 56,* 243–256.

Slotkin, R. (1992). *Gunfighter nation: The myth of the frontier in twentieth century America.* New York: Athenaeum.

Suvin, D. (2005). Estrangement and cognition. In J. Gunn & M. Candelaria (Eds.), *Speculations on speculation* (pp. 23–35). Lanham, MD: Scarecrow Press.

Wagner, J. & Lundeen, J. (1998). *Deep space and sacred time:* Star Trek *in the American mythos.* Westport, CT: Praeger.

Wilson, G. (1986). *Narration in light.* Baltimore, MD: Johns Hopkins University Press.

Worster, D. (1991). Beyond the agrarian myth. In P. Limerick, C. Milner, & C. Rankin (Eds.), *Trails: Toward a new western history* (pp. 3–25). Lawrence: University of Kansas Press.

• CHAPTER FIVE •

# Neocolonialism, Technology, and Myth in the *Stargate* Universe

*Scott Simpson and Jessica Sheffield*[*]

Science fiction is an existential metaphor that allows us to tell stories about a human condition. Isaac Asimov once said, "Individual science fiction stories may seem as trivial as ever to the blinded critics and philosophers of today, but the core of science fiction, its essence, has become crucial to our salvation, if we are to be saved at all."

— Douglas Anders (*Stargate SG-1: 200)*

Myths have been part of human culture and civilization throughout history. They have served as religion, as the underlying narrative by which groups constitute themselves, and as the fears and hopes of cultures. Today, science fiction and fantasy take part in reconstituting old myths and making new ones, often through electronic mediated forms. Science fiction, especially, often serves as a lens through which to examine current and historical events and issues (Consalvo 2004; Ott & Aoki, 2001, 2004). The quotation that opens this chapter showcases science fiction's place as modern mythology: in *Stargate SG-1*'s 200th episode, a self-referential parody of many of the show's recurring themes, the writers close by citing renowned science fiction writer Isaac Asimov to justify their own places as narrators and critics of the human condition.

In this chapter, we examine the *Stargate* franchise, among the most successful science fiction series ever aired on television. It is

[*] Scott Simpson (M.A., Colorado State University) is a Ph.D. student at the Pennsylvania State University. Jessica Sheffield (M.A., The Pennsylvania State University) is a Ph.D. student at the Pennsylvania State University.

currently constituted by two cable television shows: *Stargate SG-1* and *Stargate Atlantis*. The first show, based on the 1994 *Stargate* film, became the longest continuously running science fiction show in the summer of 2006 before being cancelled during its tenth season by the Sci-Fi Channel. *Stargate Atlantis*, the second spin-off, debuted in 2004 and, as of this writing, was still on the air.[1] Through more than 200 episodes, the two shows have addressed issues of religion, societal control, ethics, technology, and war. Along the way, the shows have drawn upon the myths of various Earth cultures, building a universe colonized by the gods and demons of Earth mythology. We contend that while the shows present a somewhat utopian vision of the world coming together to face insurmountable odds, they also weave in dominant ideologies of colonialism and a belief that technology will ultimately save humanity.

The premise of *Stargate SG-1* is a familiar one to science fiction viewers: an intrepid team of explorers visits other worlds, making contact with races both familiar and alien. The show picks up one year after the film, in which a device—the Stargate—is discovered that allows instantaneous travel to other worlds in the Milky Way galaxy, simply by activating it and stepping through. The Stargate is controlled by the military, but it is a civilian scientist named Daniel Jackson who discovers how to operate it and is part of the first team to travel through it. In the film, upon arriving on their first alien world, the team liberates the local population from an evil overlord, an alien posing as Ra, the Egyptian sun god. This theme is picked up in the *SG-1* narrative: Ra is part of a race of parasites called the Goa'uld, aliens who possess human bodies and pose as gods from Earth's mythologies to keep their human slaves under control. Because of their parasitic nature, the Goa'uld do not invent their own technology but incorporate the technology of those they have taken over or conquered into previously found or conquered technology. Throughout the show, SG-1 and the other members of Stargate Command (the SGC) battle with Goa'uld posing as Apophis (Egyptian demon of darkness and chaos), Hathor (goddess of fertility), Anubis (god of the underworld), Nirrti (Hindu goddess of death), and many others, fighting to free their human slaves and to keep Earth from suffering a similar fate. While other alien races are introduced as both allies (the Tok'ra, the Asgard) and enemies (the Replicators, the Ori) of Earth, it is the war with the Goa'uld that serves as the primary narrative thread.

Late in the sixth season of *SG-1*, members of the SGC begin the search for a legendary "lost city," which they hope will contain weapons to help them in their fight. The city in question is, of course, Atlantis. Its discovery leads to the spin-off show *Stargate Atlantis*, in which an expedition consisting of both military and civilian members travels to the Pegasus galaxy to explore Atlantis—and, through its Stargate, other worlds in that galaxy. The initial journey is one way, as travel to another galaxy takes an enormous amount of power, and there is no way to power the Atlantis Stargate to send the expedition back to Earth if trouble arises. The theme of the show then becomes the search for a power source in addition to the exploration of new worlds—and the struggle against a new enemy, the Wraith, a race of powerful, near-immortal aliens who feed on humans in the Pegasus galaxy. During their initial exploration of the galaxy, the Atlantis expedition accidentally awakens the Wraith from hibernation. In addition to their own peril, then, the expedition is responsible for unleashing the Wraith upon the largely primitive cultures of the Pegasus galaxy. Therefore, they must both protect their Pegasus neighbors and keep the location of Earth from the hungry aliens.

With storylines that address war, religious fanaticism, intercultural relations, gender issues, and more, the *Stargate* franchise offers a rich text for analysis. Through a postcolonial critique, we trace the intersections of religion, race, gender, and technology in the *Stargate* shows. We first give an overview of the neocolonialist perspective and then discuss the mythos of technology in the *Stargate* universe. We then move to a discussion of neocolonialism as exhibited in the show. Finally, we offer a consideration of how technology serves as the determinative factor by which neocolonialism functions in *Stargate SG-1* and *Stargate Atlantis*.

## Neocolonialism

Postcolonialism, which is "a critical perspective that primarily seeks to expose the Eurocentricism and imperialism of Western discourses," has influenced a wide range of fields in the humanities (Shome, 1996a, p. 41). Shome argues that when we engage in understandings of texts or in the production of theories, we need to place them within the larger context of neocolonialism and racism (1996a, p. 41). Echoing her 1996 essay, Shome and Hegde (2002) argue that postcolonial scholarship "theorizes the geographical, geopolitical, and historical specificities of modernities *within* which other forms of power—such as race, sexuality, culture, class, and gender—are located" (p. 253). In this chapter, we focus primarily on the intersections of religion, race, and gender as they are represented in the neocolonial narrative of the *Stargate* universe.

For Shome (1996a), there are three perspectives postcolonial critical scholars can take: discursive imperialism, hybrid and diasporic cultural identities, and postcolonial academic self-reflexivity. In this chapter we take the perspective of discursive imperialism, which "challenges the colonizing and imperialistic tendencies manifest in discursive practices of 'first world' countries in their constructions of 'third world' countries and/or racially oppressed peoples of the world" (Shome, 1996a, p. 42). Of particular interest to us is the circulation of images in popular culture. We follow Shome and Hegde (2002) in asking, "How is knowledge about 'other' cultures reproduced without affecting the global hegemonic order?" and "How are different national groups represented in depictions of globalization?" (p. 265) in the *Stargate* franchise.

One focus of postcolonial critics is the neocolonialism of nations such as the United States and the United Kingdom in their "representations of subjects of developing countries and racially oppressed groups...as an 'other'—racially inferior and hence open to subjection by (white) Western discursive practices" (Shome, 1996a, p. 42). These discursive practices are the primary means of imperialism today in an increasingly globalized world. Rather than through conquering territory, subjugation occurs discursively through representations of the "first world" (positive representation) and the "developing world" (negative representation). These representations are further amplified by the media systems of the global economy, in which the United States in particular produces the vast majority of the world's media products. Our goal is to follow Shome and Hegde (2002) in providing "historical and international depth to the understanding of cultural power" (p. 252).

Through the contemporary texts *Stargate SG-1* and *Stargate Atlantis*, we engage in a neocolonial critique that seeks to understand the discursive constructions of "others" in relation to dominant hegemonic discourses about the world. Such an understanding requires the critic to attend not just to the material world but to those signs, symbols, and representations that occur within the everyday and popular culture (Parameswaran, 2002). Buescher and Ono (1996) have explored neocolonial rhetoric specifically in Walt Disney's animated feature *Pocahontas* (1995), arguing that the film is a neocolonialist text because it "transforms colonialism into a benevolent ideology of good will, proto-environmentalism, proto-feminism, and cross cultural tolerance, a soothing tonic with which to heal public social ills" (p. 129). Most important for understanding a neocolonialist text is recognizing that current colonialist practices "operate under the conscious threshold of the contemporary imagination," making neocolonialism "contemporary culture's willful blindness to the historical legacy of colonialism enacted in the present" (p. 130). For Buescher and Ono, neocolonialism is tied not only to the past but to how the present is (re)imagined in the past through the retelling of a colonial narrative. We argue that the neocolonial narrative in *Stargate* shares many of the attributes that Buescher and Ono describe, but that the myths of the past are (re)imagined in the present to justify contemporary neocolonial practices. Contemporary ideologies of feminism, economics, militarism, and cross-cultural tolerance are deployed in the alternate reality of the *Stargate* universe to mask current neocolonial rhetorics and representations.

One of the main themes of the neocolonial narrative in the *Stargate* universe is its reliance on myth. Carey (1988) reminds us that myths are "persistent practices" that "sediment the world together" (p. 15). Myths, then, help us to make sense of the world we live in. Some myths are more persistent than others; for example, Rushing and Frentz (1995) lay out colonialist practices as they are reflected in the hunter myth in the United States. They argue that the frontier myth rooted in the Indian hunt is one of America's most enduring myths. This myth evolves through three phases—the Indian, the frontier, and the technological hunter—with elements of all making their way into the myths of today. As the myth of the hunter progresses a colonialist rhetoric is developed that displaces indigenous peoples with a Eurocentric view of progress and an enlightened (rather than "primitive") self. Such a myth, however, also creates a tension between the self and other, and between technological determinism and social interjection in the real world, as we discuss later in this chapter.

The neocolonial narrative in the *Stargate* universe is intimately tied to ideas about technology and its role in human affairs. With advances in technology come new myths; for example, scholars have examined the mythos of the electronic revolution (Carey, 1989, 1992;

Mosco, 2004). Mosco (2004) however, is quick to remind us to pay attention to the material world and the consequences that technology and its attendant myths have for society. Technology is not only the driving force behind the *Stargate* storylines but also the linchpin that highlights representations of material colonization and imperialism. We find, for example, that the Goa'uld did not build the Stargate system but rather use it for their own colonial practices. The Stargates were built by a race known as the Ancients, who, we learn, are the ancestors of the human race. The story is reminiscent of early archeological explanations of ancient monuments such as the Pyramids in Egypt and the Americas, the statues on Easter Island, and myths of a lost white race in Africa who built a city of gold: in the eyes of Europeans and Americans, indigenous peoples could not have built the monuments that they live near. Stories like these underscore the importance of technology as part of the neocolonial mythos of the *Stargate* universe.

## Technological Mythos in *SG-1* and *Atlantis*

Science fiction has long showcased both humanity's enthrallment with and fears about technology. While Jules Verne took readers on an exciting voyage in *Journey to the Center of the Earth* (1864), H. G. Wells and Mary Shelley explored the darker sides of scientific research in *The Island of Dr. Moreau* (1896) and *Frankenstein* (1818). From its roots in European works such as these, the genre "grew up in the United States alongside technological futurism, and shares its fascination with future gadgetry, if not all of its enthusiasm about the future" (Braine, 1994, p. 6). Similar to Whitt's analysis of Cyborg (this volume), encounters with new technologies are primary storytelling devices in the *Stargate* shows, and they complicate the mythos of the *Stargate* universe. *Stargate* episodes often feature problems of and with technology, which then require yet more technology by way of solution—what might be called in a Heideggerian sense the "technological understanding of being" (Dreyfus, 1995, p. 47). Thus, as presented in the show, technology both expands and limits our understanding of the world and our place in it.

In this section, we offer three perspectives for understanding the technological mythos of the *Stargate* universe: technological utopianism, technological utilitarianism, and technological determinism. These three perspectives intertwine to support the neocolonial narrative presented by the show's authors—indeed, such colonization would not be possible without technology. We discuss this relationship further in this chapter's conclusion.

## Technological Utopianism

Technological utopianism in the *Stargate* universe is most cogently articulated in the form of Atlantis, the mythical city of the ancestors of humanity. The "City of the Ancestors," as natives of the Pegasus galaxy call it, is beautiful and mysterious—abandoned for thousands of years, it comes to life when the expedition from Earth steps through the gate. We experience the architecturally stunning city through their eyes, with graceful spires rising hundreds of feet into the air. Many of the windows are made of stained glass, giving the impression that one is standing in a cathedral. The stairs to the Stargate are lit from beneath and inscribed with Ancient writing, and even the "drones," weapons with hugely destructive power used in defense of the city, are glowing works of art. It is also a technological wonder: it survived ten thousand years at the bottom of the sea and, were it fully powered, would be capable of traveling through space.

Mythological visions of technological utopia or dystopia are certainly not the exclusive domain of television. Technological utopianism drew from the eighteenth-century ideal of the progressive idea of history, arguing that a perfect future could be made possible by technological innovation (Braine, 1994; Marx, 1987). In this view, technology was society's engine, not merely its byproduct. Writing about utopian novels of the late nineteenth century such as *A Connecticut Yankee in King Arthur's Court* by Mark Twain and *A Traveler from Altruria* by William Dean Howell, Kasson (1999) argues that utopianism "functioned as 'a speculative myth,' a mode of projecting a vision of social possibilities at a time when Americans felt themselves at a turning point in history, but desperately in need of a sense of direction" (pp. 190–191). Each of the novels Kasson analyzes, in its own way, examined "the meaning and relevance of egalitarian ideals in a technological society, and the possibility of achieving a republican utopia" (p. 190). These books, Kasson contends, appealed to Americans who were increasingly troubled by an apparent conflict over American society and values.

Science fiction, in particular, is an excellent vehicle for providing commentary on the ability of technology to help bring about a better future. Writing about *Star Trek: The Next Generation*, Braine (1994) notes that its recurring tensions between "adventure and security, control and disaster, optimism and anxiety, the futuristic and the old fashioned, are in essence the dynamics of American technological utopianism, as it evolved from the later nineteenth century to the present" (p. 4). The *Stargate* universe makes sense of attendant anxieties over technology by subsuming them beneath the social interaction between characters and cultures. This social interaction is expressed as a neocolonial narrative simultaneously relieving anxieties over technology and contemporary colonial practices. *Stargate SG-1* and *Atlantis* function to support the status quo—that is, nothing needs to be changed because everything is working as it should.

As alternate reality science fiction, the *Stargate* universe invites its audience to imagine a different present. This imagined universe seeks to quell questions about contemporary technological and neocolonial practices in the real world. Ott and Aoki (2001) demonstrate how "the utopian appeals of [*Star Trek:*] *The Next Generation* affirm and recenter White heterosexual masculinity in U.S. popular culture" (p. 410). They find that "visions of the future will always serve particular political interests and relations of power" (p. 410). Although the focus for Ott and Aoki is on utopian discourses as a prescriptive function within the popular imagination, alternate reality programs function slightly differently. Like *Star Trek: The Next Generation* and many other science fiction programs, the two *Stargate* series refine and rearticulate "the American fantasy of the technological future, with all its attendant anxieties" (Braine, 1994, p. 3).

This is not to say, of course, that the shows consistently paint the technological future as utopian. In fact, many individual episodes deal with deep-seated unease about the use of technology, particularly in cases when the humans are uninformed about the consequences of that use. For example, in a *Stargate Atlantis* episode called "Trinity" (a reference to the United States' nuclear program), astrophysicist Rodney McKay is responsible for the death of a colleague and the destruction of five-sixths of a solar system when he attempts to repair an ancient power system. Episodes such as "Trinity" and *SG-1*'s "Absolute Power," in which Daniel Jackson is shown a vision of the terrible destruction he will wreak on a helpless Earth if he is given the technological knowledge he thinks he needs to defeat the Goa'uld, display a dystopian vision of technology. These episodes are necessary in order to reinforce the overarching utopian paradigm: at the end, the characters have learned their lessons about the negative use of technology, and progress toward the perfect technological future may proceed.

## Technological Utilitarianism

Another technological perspective we present is technological utilitarianism. In the first season *SG-1* episode "The Nox," the SG-1 team encounters a seemingly primitive race who are much more advanced then they appear. Before the team leaves, though, we are given a glimpse into the politics of the Stargate program. The team is reminded by the U.S. Secretary of Defense that their mission is more than mere exploration—it is meant to produce technological advances for Earth:

> Secretary: Do you have any idea what's out there?
> O'Neill: No, sir—that would be why we're going.
> Secretary: I'm not sure—Colonel, is it? Because to be perfectly frank, this administration is not satisfied with the current progress of the Stargate program.
> Carter: Begging the secretary's pardon, sir, but we've already visited nineteen separate worlds.
> Hammond: I believe he is referring specifically to the volume of technology being returned on your planetary missions.
> Secretary: The president and Joint Chiefs were under the impression that the SG teams would bring back superior technologies.
> Daniel: I'm sorry, I thought we were explorers.
> Secretary: Oh, you are, Dr. Jackson—but even Marco Polo, when he came back from the Far East, brought more then just a few exotic spices.

In this exchange, we see the emphasis on returning technology to Earth for use by the United States government. No mention is made of evaluating such technology or of different contexts in which the technology might be used. The assumption is simply that advanced technology will be of benefit to the human race and that exploration for its own sake is not a desirable goal. In the scene, Daniel Jackson resists this imperative, but it is at the heart of all exploration by SG-1.

Through science and technology, then, the survival of humanity as a species is guaranteed, even if it requires the sacrifice of a few. Yet it is not human innovation alone that will save the day. Rather, the characters on the show rely on technology from more advanced races to augment their own limited resources. The mission of both the SG-1 and Atlantis teams is to bring back advanced technology from more evolved races (such as the "Ancients," the builders of the Stargates and ancestors of humanity). Such artifacts then become the domain of the shows' scientists, who study the devices and adapt them to human use. This technological mythos of the show is one of *utility*: technological artifacts are considered from the perspective of how they are made and how they can be used, not whether or in what circumstances that use should occur (Winner, 1986). This ends-based perspective values advanced technology above and beyond humanity's capacity to invent it for ourselves, creating a worldview in which the answer to almost any problem is to find sufficiently advanced technology and put it to use.

In some episodes, technology allows the humans of Earth to quite literally rewrite history. For example, in the *SG-1* two-part episode "Moebius," the SG-1 team travels back in time to find a Zero Point Module, or ZPM, a source of nearly limitless power. When their actions change the timeline so that the future is fundamentally changed, they must again travel into the past to undo their actions and set things right. In "Before I Sleep," an *Atlantis* episode, time travel allows the leader of the expedition to save the lives of the entire expedition when she is thrown backward in time and has the opportunity to forestall the disaster that will kill them before it happens. Technology, then, becomes not just a promise of the future but a means by which the past can be rewritten.

## Technological Determinism

A third articulation of technology in the *Stargate* universe is that of technological determinism. Viewed through this perspective, technology is an omnipresent, undeniable force that shapes the course of human history (McLuhan, 1962, 1964). In the shows, more technologically advanced societies such as the Nox and the Tollan are depicted as reluctant to share their technology with the humans of Earth because they fear what the use of that technology will bring about. In a scene from a first-season *SG-1* episode, "Enigma," Narim, a member of the Tollan race, tells Samantha Carter, a member of SG-1, why the Tollan are opposed to sharing advanced technology with Earth:

> Narim: The nearest planet in our solar system was called Surita. When we began to explore space, we learned she was inhabited.

Carter: Did you make contact?
Narim: Yes, when we thought they were sufficiently advanced. They were on a level very similar to yours.
Carter: So what happened?
Narim: We offered them a device to produce unlimited productive energy. And they used it to make war.
Carter: How bad was it?
Narim: In one rotation of our planet, they had destroyed theirs.
Carter: One day?
Narim: The destruction shifted our planet's orbit point 3 tekanna, enough to begin a chain of events that made our world unstable.
Carter: So that would explain why [the Tollan are] so afraid to give a society like ours any of your technology.
Narim: Yes.

The series often explicitly shows ambivalence on the part of "older" races toward advancing humanity beyond their current technological capability. This is, in many ways, a storytelling device: if the humans of Earth came to rely too much on outside help, the show would quickly become dull.

By acting as a cultural forum, "television fictions that present a technological future may address immediate concerns about the impacts of individual technologies, but they do not question the technological society itself" (Banks & Tankel, 1990, p. 25). Although some contemporary science fiction shows (such as the reimagined *Battlestar Galactica*) do take as their subject matter the problems and perils of technology, the *Stargate* shows typically limit such inquiry to small, episode-delimited storylines. The result is that, while human courage and innovation are often the central theme of episodes, they are subsumed under a pervasive technological determinism that subordinates the natural world (Ellul, 1967).

Although it is interaction with technology that most often causes problems for the humans, they are sometimes unaware that their actions have set in motion technological events. For example, in the premiere episode of *Stargate Atlantis*, Colonel John Sheppard finds in the Athosian city of the ancestors a locket that had belonged to Teyla, an Athosian native of the Pegasus galaxy, as a child. Because he is a carrier of the ATA gene (Ancient Technology Activation gene, which some humans have inherited from their Ancient ancestors), he unknowingly activates it by touch, and it acts as a beacon that draws the Wraith to them on subsequent missions. Technology, then, becomes the determinative factor in the narrative of the series.

Disruptions in technology must often be solved through the application of other technology. One example is in the first season *Stargate*

*Atlantis* episode “Hide and Seek,” in which a child’s exploration of the city results in the failure of a crucial piece of equipment—a device containing a destructive alien force, an indeterminate cloud that drains energy from machines and people. Freed of its prison, the alien moves throughout the city, wreaking havoc on the expedition’s power reactors and other technology, even killing humans along the way. Only by luring it through the Stargate, using a reactor as bait (carried by Rodney McKay, who wears an Ancient-developed personal shield to prevent the alien from draining his energy and killing him) can the expedition be rid of the cloud. The metaphor is difficult to untangle: the technology imprisoning the alien fails, freeing it to disrupt other technology, which forces the humans to employ yet more technology in defeating it.

The Replicators are the most compelling articulation of technological determinism in the *Stargate* universe. Mindless, insectlike robots that devour everything in their path in order to create more of themselves, the Replicators threaten to destroy the whole galaxy if the humans of Earth cannot stop them. Like the Borg in the *Star Trek* universe, the Replicators “threaten humanoid hegemony because...they embody a purely technological imperative that does not serve humans, or even acknowledge them” (Braine, 1994, pp. 10–11). They swarm across the galaxy, threatening even the technologically superior Asgard. In fact, it is the Asgard’s technological prowess that threatens to doom them, as their weapons are powerless against the Replicators: it is the humans, with their technologically inferior projectile weapons, who are ultimately able to defeat the swarm of killer robots.

## The Technological Mythos of *Stargate*

In their representation of technology, the *Stargate* shows draw upon many of the conventions of science fiction: technology as a tool, as a benevolent force, or as an unstoppable power. Audiences are surely used to seeing such views of technology: after all, it is *science* fiction. Yet technology is never value neutral (Carey, 1983). It can promote certain ideologies and obscure others, making a bold statement about the universe the show is set within. According to Banks and Tankel (1990), for example, the *Star Trek* universe offers audiences “representations of the future in terms of science, technology, and social progress that (1) reinforce contemporary hierarchies, social relations, and patterns of interaction and therefore (2) allay contemporary fears that technology will alter social relations beyond recognition" (p. 26). Unlike *Star Trek*, which portrays an imagined future, the *Stargate* shows operate within a reimagined *present*; therefore, we must be particularly attentive to the ways in which technology is deployed to bolster or hinder certain ideological viewpoints.

Television fictions are, according to Banks and Tankel, "symbolic offerings that naturalize the role of technology in cultural mythmaking" (p. 24). In the *Stargate* universe, the search for advanced technology is the central storytelling device of the two shows. The focus on technology, as outlined earlier, is manifested through three major perspectives: utilitarianism, utopianism, and determinism. Even when one paradigm is questioned in the show, such as when the Tollan question the wisdom of sharing their technology with Earth, another takes its place: utilitarianism is replaced with determinism, utopianism with utilitarianism, and so on. The tensions among these perspectives provide a lens for examining neocolonialist practice in the shows.

## The Stargate as Neocolonial Practice

The purpose of the Stargate within the neocolonial narrative is to refigure both time and space in a reimagining of Earth's past. The Stargate collapses space, allowing for the instantaneous movement of people and goods throughout the universe. As long as a gate exists on a world and you have the sequence of symbols necessary to "dial" it, then you are able to get to that world. The capability to move people, goods, and weapons instantaneously is a colonist's dream. This ability is a metaphor for freedom within the *Stargate* universe, as long as it is used for the right reasons. These reasons are highlighted by a tension between malignant colonization (the Goa'uld) and benign colonization (the SGC and its teams).

The Goa'uld use the Stargate to conquer other worlds, enslave their people, and expand the territory of the system lords. As discussed earlier, the Goa'uld assimilate the technology of worlds they subjugate, relying particularly on technology developed by the Ancients (such as the Stargates themselves). The colonizing practices of the Goa'uld are represented as destructive and dominating: humans must bow to the "god" or die (a storyline that is repeated later in the series with the Ori, who seek to impose their religion on the inhabitants of the Milky Way). As "gods," the Goa'uld recall the absolute monarchs of medieval Europe, living in luxury while their enslaved subjects toil to provide them with the matériel and technology needed to maintain their power.

In contrast, the SG-1 team goes through the Stargate in "friendship," rather than as conquerors. The SGC's use of the Stargate is not portrayed as parasitic like that of the Goa'uld, even though the humans too are looking for new technology to incorporate into their existing arsenal. Rather, the SGC's actions are framed as necessary for the survival of Earth. In order to compete with the technological superiority of the Goa'uld, SG-1 *must* explore or face annihilation. Ironically, Earth's greatest weapon against the Goa'uld is discovered not on an alien planet but under the ice of Antarctica: an Ancient outpost with an arsenal of "drones" similar to those found on Atlantis.

The SGC also looks to set up trade relations with other planets, either for technology or for natural resources. Capitalism, then, becomes the basic economic system that all cultures in the universe understand: even in the Pegasus galaxy, the members of the Atlantis expedition set out almost immediately upon their arrival to find allies for the purposes of trading. However, this system is not without its criticisms. For example, the SG-1 team is often looking for naquada—a resource in the *Stargate* universe from which the Stargates are made and which is used to produce a great deal of energy—and is willing to trade medicine or other supplies for it to alien planets. Naquada, however, must be mined, and no mention of the mining practices or what alien species actually get in return for their efforts is ever made.

This narrative closely parallels the age of colonization on Earth, from the late fifteenth-century exploration of the New World through the age of empire in the nineteenth century. Colonizers initially sailed off to new lands for the purposes of trade but quickly turned their attention to the mining of gold and other resources. In some cases there may have been an initial trade treaty that quickly devolved into the outright subjugation, displacement, and killing of the indigenous population. Because no depth is given to trade negotiations, the audience is invited to believe that a win-win situation is created between the SGC and other planets. Furthermore, the SGC can be seen as benign because they are not subjugating a planet like the Goa'uld. This is the central contrast shown between the harsh colonial practices of the Goa'uld and the benevolent neocolonialism of the

SGC. Yet, as we argue below, the colonial practices of the SGC are, in their own way, ideologically charged.

Although there are many aspects of the *Stargate* shows that would be appropriate to this particular critique, we focus here on three: religion, race, and gender. We examine these themes, their mythic undertones, and their relationships with each of the technological perspectives discussed earlier.

## Religion

The Stargate is not just a piece of technology for moving people from point A to point B in the universe: it also collapses time in a figurative sense, as many of the worlds populated by humans have not advanced technologically. This allows the writers of the show to redeploy the myths of the past in the show's storylines. For example, SG-1 visits a planet that is protected by the Asgard, a benevolent alien species who take on the personae of ancient Norse gods. However, this world is in the same state of technological stagnation as the worlds of the Goa'uld. This seems strange, since SG-1 discovers that the Asgard will reveal their true form to these people once they reach an appropriate stage of development. Presumably, since the Scandinavian peoples of Earth were able to advance, this planet should be at a similar stage; the disparity is perplexing.

The explanation for this apparent discrepancy can be found in an examination of religion in the *Stargate* universe. The Asgard's technology renders them godlike to primitive humans, and the small gray aliens use holographic technology to impersonate human gods and instruct their followers. Although the Asgard are benevolent guardians, their religious practices nonetheless serve to stifle technological innovation on the planets under their protection, albeit in a less socially destructive way than is evident on Goa'uld worlds. Through depiction of alien religions as based on "false gods," the show is able to recenter white Christian masculinity, despite the fact that the Christian God is almost never mentioned in the shows. Gods that *are* mentioned are framed as "false" or explained away as alien races. In framing gods this way, the show is able to depict their followers as backward and in need of enlightenment as brought by SG-1, who often unmask the "magic" of the false gods as technology. This unmasking of false gods implies an alternative (monotheism, presumably Christianity) and serves as a subtle form of religious imperialism. SG-1 explains that worship of the false gods makes people slaves, not righteous followers. Enlightenment is not just through debunking "magic" but also through bringing an ideology of individualism to other worlds. This individualism is framed in terms of both an American and a Christian sense of identity: the desire to set others on the "correct" path. In the representation of contemporary U.S. society in the *Stargate* shows, ancient mythologies and pantheism are understood through the lens of Christianity, in which the intertwining of Christendom, economics, and democracy pervade U.S. nationalistic and patriotic discourse.

The religions represented on the shows also reflect colonialist practices on Earth. For example, many of the Goa'uld are given the names of ancient Mesopotamian, Egyptian, Chinese, and Celtic gods: Amaterasu is the sun goddess from Japanese mythology, Camulus is from Celtic mythology, and Zipacna is from Mayan mythology, to name a few. Throughout the series at least fifty-five Goa'uld are seen or referred to by other characters. Most were the gods of people who were subjugated and colonized at some point in their history by European colonial powers, in which religious missionary work and conversion to Christianity played a large role. Polytheistic religion thus stands in for barbarism in the narrative of the show; while Christianity is not explicitly offered as an alternative, the implication is that peoples visited by SG-1, unchained from their "false" faiths, are now free to follow the "true" path. Furthermore, when the Ori are introduced and Dr. Jackson links them to myths of the devil on Earth, by implication SG-1 is fighting on the side of the "true" faith.

Although not explicitly articulated in the show as Christianity, the highest form of spirituality is expressed in a utopian vision of technology. For example, the act of walking through the Stargate can be seen as a sublime technological baptism. After the initial rush that follows the activation of the gate, the event horizon of the wormhole is depicted as calm, clear water. The point of view of stepping through the gate is that of the SGC or Atlantis teams: we follow them, not their enemies, across the threshold. First-time travelers are often depicted as awestruck and hesitant before entering the gate. These travelers' trepidation exemplifies a lack of faith that is overcome by taking the first step toward a new birth on a different planet. Thus, the same technology that allows SG-1 to "liberate" alien worlds from "false gods" also provides its own connection to the divine.

Even more illustrative is the "Ascension" storyline, in which we learn that the Ancients advanced so far, both technologically and socially, that many of them shed their physical bodies and became beings of pure energy with extensive powers. These beings are composed of white light and, although not strictly humanoid in appearance, nonetheless evoke Christian angels. Furthermore, since the city of Atlantis represents the pinnacle of Ancient civilization, its resemblance to a cathedral visually reinforces the connection between technological advancement and a higher state of spiritual being.

Ascension is represented in the shows as the utmost spiritual and technological condition to which someone can aspire. However, those who do ascend must adhere to a strict code of ethics regarding interference with "lesser" beings. This code causes consternation among the heroes of the show, who, after all, spend much of their time interfering with other cultures. For example, an ascended being named Oma Desala (translated by Daniel Jackson as "Mother Nature") helps Jackson ascend after he is exposed to a lethal dose of radiation, but he later chooses to return to corporeal form after becoming frustrated by limitations on his powers, especially when his friends were in danger. John Sheppard, military leader of the Atlantis expedition, is given the chance to ascend in a season two episode of *Stargate Atlantis*; he too chooses to remain bodily in the world, saying that he is "not ready" for ascension. Ascension also has its dark side: the Ori, villains of seasons nine and beyond of *Stargate SG-1*, are themselves ascended beings who, like the Goa'uld, have chosen to use their powers for domination. Despite being represented as a divine process, ascension also thus serves as a rhetorical examination of the perils of absolute power.

## Race

The concept of race would, at first, seem to be effaced in the *Stargate* universe. Various SGC teams are racially mixed, as is the Atlantis expedition, and the characters generally do not explicitly discuss Earth's racial issues. Contemporary depictions of race are transposed instead onto off-world humans and nonhuman aliens. The main depiction of race in *Stargate SG-1* and *Stargate Atlantis* is an amalgamation of racial stereotypes based in a U.S. and Eurocentric view, often mixed with religious overtones.

Teal'c, the physically imposing alien member of SG-1, is the noble savage/warrior and is an amalgamation of African, African American, and Native American stereotypes. He is played by an African American actor. Teal'c is a member of the Jaffa race, who serve the Goa'uld both as warriors and as incubators for larval Goa'uld symbiotes. Stereotypically, the Jaffa are a tribal race whose culture and practices seem wholly different from Anglo society. They—and Teal'c—are strong, taciturn, and aggressive. If something is not working, Teal'c is most likely to hit it or shoot it when all else fails.

Before becoming a member of SG-1, Teal'c served as the commander of Apophis' armies. However, in the pilot episode, "Children of the Gods," Teal'c betrays Apophis to help SG-1 free prisoners (including themselves) who have been sentenced to death. Despite this action, as someone who recently served the Goa'uld, Teal'c is not immediately trusted at the SGC. Upon arriving through the Stargate, though, Teal'c relinquishes his weapon in a symbolic gesture of surrender. In this case, Teal'c immediately agrees to help the Earth in its fight with the Goa'uld, giving up his family and status immediately and apparently without second thoughts. The only reason that he gives is that he believes the SGC can actually defeat the Goa'uld. Later, we find that there are rumblings of rebellion within the Jaffa people, but they need the help of the Earth to achieve their freedom. This is part of the neocolonial narrative: native people cannot liberate themselves without the enlightenment and help of dominant white society.

In the *Stargate Atlantis* pilot episode, "Rising," the Atlantis team encounters off-worlders—the Athosians—in friendship, rather than as enemies. As explained earlier, Atlantis expedition members accidentally reawaken the Wraith, unleashing them on the Pegasus galaxy in general and the Athosians in particular. Despite the fact that the Wraith killed or captured some of her people, after the Atlantis expedition rescues the remaining Athosians, Teyla tells Sheppard, "You have earned both my friendship and that of my people. With our help, you will make many more friends." Teyla, then, is stereotypically represented as the "good" native. She is willing to learn about other people and is open to their ways. Prior to joining the Atlantis expedition in the city, her people were nomadic and dwelled in the forest, revering both the earth and the Ancestors, their term for the Ancients who built the Stargate. Ancestor reverence and animism are elements of Native American mythology, and Teyla herself can be seen as an amalgamation of the Pocahontas and Sacagawea myths of Earth—a female native who helps the intrepid white explorers out of friendship.

Language is one mechanism for setting apart aliens as other. Strangely, all humans in both the Milky Way and the Pegasus galaxies speak American English. Other science fiction shows have at least made plausible excuses for English being used on a show (*Star Trek,* for instance, had a universal translator), whereas the *Stargate* shows offer no canonical explanation as to why the characters can instantly communicate with all the humans they encounter. While this is largely a matter of practicality, as the show would be difficult to produce if writers had to invent both a new language and a new method for overcoming the language barrier every week, English can also be seen as a marker of difference and nationalism, particularly in *SG-1.* Non-human aliens such as the Unas, primitive beings who were the first hosts to the Goa'uld, speak no English; aliens such as Teal'c, who are a mix of human and Goa'uld, speak both Goa'uld and careful English; human off-worlders speak colloquial English and rarely any other language. Breakdowns in communication, which sometimes occur, serve as a pretext for action by the SG-1 team and mark aliens as other. This "othering" process is a foundational part of a neocolonialist narrative—we need to make them more like us by teaching them our language. When "good" aliens such as Teal'c and Teyla speak English, they are still marked as different. Both characters almost never use contractions; there is no reason for this other than to set them apart as different from the other characters and mark them as "alien." Teal'c and Teyla, although from offworld, have clearly interacted with other people and aliens who do use contractions, but because they are close friends with the humans from Earth their use of the English language has to be more than perfect. In their perfection, they become exoticized and marked as different.

A utilitarian view of technology would seem to offer ways to close the gap between dominant culture and other. From this perspective, technology could be used in ways that minimize difference—yet it is all too often used to bolster dominant ideology. For example, in the episode "Affinity," the development of a immune-boosting drug called tretonin means that Teal'c no longer needs to incubate a Goa'uld symbiote to survive and can live as a human. However, in addition to his non-white physical appearance, Teal'c is still marked as other by the Jaffa tattoo on his forehead. Still, he wishes to become part of Earth society and so often dons a hat and leaves the SGC, eventually getting his own apartment off base. He develops a close relationship with his neighbor Krista, a white woman, which implies that the two of them have affection for one another. Such relationships within a neocolonial narrative, however, cannot be consummated. Krista is already dating a man who is abusive and a drunk. Teal'c teaches Krista a form of martial arts that Krista then uses to kill her boyfriend during an argument. Teal'c is blamed for the death and must return to his quarters at the SGC. The episode shows the incompatibility of racial mixing: concepts such as multiculturalism are eschewed for a rigid racial hierarchy in which people must be put in their place. Teal'c is figuratively put in his place by moving back into the SGC even though his neighbors like him. The episode supports a segregationist attitude masked behind a pragmatic approach to dealing with people who are different.

A more radical use of technology in the shows indicates an even more blatant colonial practice: that of transforming and subsuming the "other" completely. The humans of *Stargate Atlantis* attempt to defeat the Wraith by creating a retrovirus that eliminates the alien portion of their DNA, rendering the terrifying Wraith into harmless, amnesiac humans. The memory component is symbolically important: not only does the virus refigure physical appearance; it also removes any recollection of life as a Wraith, allowing the humans to rewrite memory as they see fit.

Unlike tretonin, which can be viewed as a tool of liberation (freeing the Jaffa from the need to incubate Goa'uld symbiotes in order to survive), the Wraith-human retrovirus is a biological weapon whose result would arguably be genocide. Yet the decision not to use it comes not as a product of ethical discussion but from the retrovirus's failure. When their test subject, "Michael," begins to remember his former life, the humans realize the retrovirus will not be an effective neutralizer. Michael reverts back to a Wraith and flees to his "hive" but subsequent episodes reveal that as a result of his transformation, he is not fully accepted back into Wraith society and is caught between the two worlds. The leader of his "hive" demands the retrovirus from the Atlantis expedition so that they can attack other Wraith with it and, having turned them temporarily into humans, feed on them. The humans are then finally forced to confront the ethical ramifications of their invention: is it right to allow the Wraith to use the retrovirus, given that their victims will—at least temporarily—be human and therefore worthy of concern?

Faced with their own destruction, the humans agree to hand over the retrovirus. However, we will never know the outcome of that decision, for the Wraith betray the uneasy truce and, at the end of the storyline, the expedition members are forced to (presumably) kill Michael and dozens of other transformed-but-reverting Wraith in order to prevent their exodus. This action guarantees both the safety of Atlantis and the resolution of their ethical dilemma. The retrovirus may have been a mistake, but all traces of its use have been obliterated. Once again, the humans have rewritten memory to their own ends, assuring their survival.

## Gender

Representations of gender have played an important role in colonialist practices focusing on the erotic and exotic "other" (Said, 1994). The neocolonial narrative of the *Stargate* universe deploys gender in at least two ways to support the dominant ideology of U.S. white patriarchy. First, a mediated version of liberal feminism is deployed as a contrast to "other" cultures' sexist practices, thereby intersecting with race. Second, gender is refracted through the lens of technological determinism, presenting women as deficient compared with men.

The Samantha Carter character is an articulation of mediated feminism, the depoliticized and appropriated form of liberal feminism in the contemporary U.S. This is a version of feminism that at first would seem to be feminist but is not. The Carter character is highly intelligent: she has a doctorate in astrophysics and is also a captain in the U.S. Air Force. By defining feminism in this way, the show maintains a gender binary and hierarchy between men and women while simultaneously allowing the show to deploy its mediated version of feminism to alleviate anxiety over current neocolonial practices. Carter, then, becomes a trope for feminism with which all other portrayals of women on the show are contrasted.

The episode "Emancipation" highlights the intersections of gender and racism in a neocolonial narrative. In this episode, the SG-1 team travels to a planet that is culturally similar to ancient Mongolian civilization, except that women must remain covered when in public. Upon their arrival, the team saves a young boy, who takes them to his village. In the interest of anthropological study, Carter agrees to wear the traditional clothing of the people. When Jack, Daniel, and Teal'c come to check on her, they stand silently with mouths agape at the way she looks. They have never seen her out of uniform in a dress, let alone a low-cut one with silk and jewels. The scene is treated somewhat humorously, but it is only when Carter is eroticized, exoticized, and feminized that she is seen as a woman rather than as just "one of the guys."

This scene simultaneously downplays both the achievements Carter has made in her life and the goals of feminism generally. Because Carter has to pass as one of the guys in her career, it is only when she wears "appropriate" female clothing that she can be seen as a woman. The clothing itself accentuates Carter's appearance as woman, and she is no longer seen as a Ph.D. holder and soldier but as a sex object. Her beauty is only evident in reference to the exotic clothing that she is wearing, which marks her as different.

Carter also represents an essentialized view of feminism in which all women are the same, and it is the job of white women to liberate women of color. In the "Emancipation" episode, after being taken captive by a rival tribe, Carter sits with one of the women who is going to be married off against her will. This woman, Nia, is the daughter of the chieftain and tries to defend him for what he does. Carter interrupts, asking, "How can you defend him when he is packing you off to marry some stranger?" Nia responds, "I have no choice. I am not free to choose." Carter states, "And you never will be until one of you says 'No'." Nia agrees with Carter: "I want to say it. Will you help me?" They then hold hands—sisters in Carter's idea of feminism.

Carter, then, is an interjection of a contemporary mediated ideology into a figurative past in which the Stargate allows for her travel to other worlds to try on other cultures and change them as she sees fit. As we noted earlier, the distance the Stargate creates in time figuratively allows audiences to view past and current practices in the world as unjust. Women of color in other parts of the world (read: not the U.S.) are in need of rescue from their men and culture. This in turn makes a colonial past appear necessary while selectively forgetting the disparities that the colonial past has created in the contemporary world.

A striking example of mediated feminism occurs in the episode "Birthright." In this episode, a radical version of feminism is deployed as a version of the Amazon Warrior myth. A group of Jaffa women have trained themselves to become warriors—going against Jaffa tradition, which dictates that only men are warriors. These women need symbiotes for themselves and their offspring and begin hunting down Jaffa males to get them. These women have been subjugated by the Goa'uld Morloc, who decreed that only male children would be allowed to live: all female children are burned alive at birth. SG-1 arrives at the behest of their leader, Ishta, who wants an alliance in order to get supplies and weapons. Ishta has been saving the female children along with a few other women. SG-1 offers to help by providing the experimental drug tretonin to the women so that they no longer need symbiotes.

In contrast to its representation as a potential liberator for the male Jaffa like Teal'c, here the tretonin drug serves two purposes in the gendered neocolonial narrative of the show. First, it is used to put the women in their place. Contrasted with Carter, the women are too radical, and although these women have the same goals as SG-1, their means of achieving them are too extremist for Teal'c and the team. Carter sympathizes with the plight of the women but cannot condone their actions. The drug, then, is a way to mitigate the rebelliousness of the women, not unlike the prescribing of antidepressants or other psychotropic drugs. Second, although appearing benign, the use of the drug shows the exploitation of those at the edges of the social order. Viewers come to learn that tretonin is experimental and has been tried on only two Jaffa in the past. The SG-1 team may be trying to help, but they are also conducting a large-scale experiment on women and girls who are at the extreme margins of their society. Women and girls of the developing world and elsewhere also tend to be marginalized and thus open to exploitative labor and medical practices.

Refracted through a deterministic lens, technology is also deployed to present women as deficient to men. Jack O'Neill as the linch pin of the gender hierarchy and gender binary is further illustrated in "Lost City Part I" and "Lost City Part II." In these episodes, O'Neill comes across a device that imbues him with the knowledge of the Ancients about where to find a powerful weapon in order to save Earth from an impending attack. Thus, even though Carter is brilliant and may be able to save the day with her knowledge, it is O'Neill who will ultimately save the world. Not only does O'Neill get the knowledge needed to save the world, but he has the ATA genetic marker in him that allows him to use the defensive systems in the Ancient outpost they find on Earth. Similarly, in *Atlantis*, it is John Sheppard who has the ATA gene, allowing him to power up and use the Ancients' technology. The show then makes a biological as well as a social argument for man's superiority to woman. The use of technology becomes a biological imperative that places men at the top of the gender hierarchy. The invention and use of technology has historically been the domain of men. The lack of the ATA gene in the main female characters shows their deficiency, just as the lack of a phallus has long been used to make the same argument. This type of narrative reproduces the dominant construction of gender in a contemporary globalized society in which technology and knowledge work is the domain of males and low-level production and information work is the domain of females. The neocolonial narrative masks the disparities between male and female workers through a technological and biological determinism.

## Conclusion

By paying attention to both a postcolonial perspective and representations of contemporary anxieties over technology, we have highlighted the ways in which technology and ideology intersect in science fiction. Carey (1983) reminds us that technology is never ideologically neutral. He argues that communications technology affects the "natural attitude" and "practical consciousness" toward space and time and often plays a role in the construction of empire, both economic and national. Schiller (1991) draws similar conclusions, contending, "Media-cultural imperialism is a subset of the *general* system of imperialism. It is not freestanding; the media-cultural component in a developed, corporate economy supports the economic objectives of the decisive industrial-financial sectors" (p. 13). Although Schiller is concerned primarily with corporate capitalism and consumerism, he highlights the intersections of the material economic structure and its relationship to mediated representations. These representations are often forced on the developing world by the developed world. To Schiller's corporate/consumer capitalism, we would add dominant ideologies of religion, race, and gender.

Within such a system, technology becomes the driving force in globalization and cultural imperialism. The anxieties over this technology and the resultant sociocultural consequences must be (re)presented through a neocolonialist narrative in which, according to Buescher and Ono (1996), "Neocolonialists depend on a history of successful colonialism, forgetfulness that colonialism continues, and the production of therapeutic stories to quell any lingering dis-ease with continuing practices of disenfranchisement" (p. 130).

*Stargate SG-1* and *Stargate Atlantis* construct an alternate vision of the present realized in a contemporary world. Fantastic technology, aliens, and worlds are contextualized as part of the myths that are at the heart of Western society. In their exploration of the universe, the characters of the *Stargate* franchise advance a Western ideology of neocolonialism; through the various crises and conflicts the characters face, the shows construct for viewers an Earth-centric universe, in which Earth is collapsed into dominant U.S. American values in which science is the dominant paradigm, technology is the resolution to human problems, and a white masculine individualism is valued above all else.

A reading such as this one challenges viewers to study what is otherwise seen as "entertainment media" from a more critical perspective. By examining what lies beneath the surface of popular science fiction, we can better understand how dominant ideologies are deployed and reinforced through mediated cultural forms and attendant material practices. *Stargate SG-1* and *Stargate Atlantis* provide us with a means toward understanding the ways in which Western colonial practices are perpetuated through a scientific and technological discourse.

## Note

1 This chapter does not address *Stargate: Infinity*, the short-lived animated series that was based on *SG-1* but not affiliated with that show's producers and is not considered official "canon" in the *Stargate* universe.

## References

Banks, J. & Tankel, J. D. (1990). Science as fiction: Technology in prime time television. *Critical Studies in Mass Communication, 7*, 24–36.

Braine, F. S. (1994). Technological utopias: The future of the *Next Generation. Film & History, 24*, 2–18.

Buescher, D.T. & Ono, K.A. (1996). Civilized colonialism: *Pocahontas* as neocolonial rhetoric. *Women's Studies in Communication, 19*, 127–153.

Carey, J. W. (1983). Technology and ideology: The case of the telegraph. *Prospects, An Annual of Cultural Studies, 8*, 303–325.

Carey, J. W. (1988). Introduction: Taking culture seriously. In J. W. Carey (Ed.), *Media, myths, and narratives* (pp. 8–18). Newbury Park, CA: Sage.

Carey, J. W. (1992). *Communication as culture*. New York: Routledge.
Consalvo, M. (2004). Borg babes, drones, and the collective: Reading gender and the body in *Star Trek*. *Women's Studies in Communication, 27*, 177–203.
Dreyfus, H. L. (1997). Heidegger on gaining a free relation to technology. In K. Shrader-Frechette & L. Westra (Eds.), *Technology and values* (pp. 41–54). Lanham, MD: Rowman & Littlefield.
Ellul, J. (1967). *The technological society*. New York: Vintage.
Kasson, J. F. (1999). *Civilizing the machine: Technology and republican values in America, 1776–1900*. New York: Viking Press.
Marx, L. (1987). Does improved technology mean progress? In S. E. Goldberg & C. R. Strain (Eds.), *Technological change and the transformation of America* (pp. 23–36). Carbondale: Southern Illinois Press.
McLuhan, M. (1962). *The Gutenberg galaxy: The making of typographic man*. Toronto: University of Toronto Press.
McLuhan, M. (1964). *Understanding media: The extensions of man*. Cambridge, MA: MIT Press.
Mosco, V. (2004). *The digital sublime*. Cambridge, MA: MIT Press.
Ott, B. & Aoki, E. (2001). Popular imagination and identity politics: Reading the future in *Star Trek: The Next Generation*. *Western Journal of Communication, 65*, 392–415.
Ott, B. & Aoki, E. (2004). Counter-imagination as interpretive practice: Futuristic fantasy and *The Fifth Element*. *Women's Studies in Communication, 27*, 149–176.
Parameswaran, R. (2002). Local culture in global media: Excavating colonial and material discourses in *National Geographic*. *Communication Theory, 12*, 287–315.
Rushing, J. & Frentz, T. (1995). *Projecting the shadow: The cyborg hero in American film*. Chicago: University of Chicago Press.
Said, E. (1994). *Orientalism*. New York: Vintage Books.
Schiller, H. (1991). Not yet the post-imperialist era. *Critical Studies in Mass Communication, 8*, 13–28.
Shome, R. (1996a). Postcolonial interventions in the rhetorical canon: An "other" view. *Communication Theory, 6*, 40–45.
Shome, R. (1996b). Race and popular cinema: The rhetorical strategies of whiteness in *City of Joy*. *Communication Quarterly, 44*, 502–518.
Shome, R., & Hegde, R. (2002). Postcolonial approaches to communication: Charting the terrain, engaging the intersections. *Communication Theory, 12*, 249–270.
Winner, L. (1986). *The whale and the reactor*. Chicago, IL: University of Chicago Press.

•CHAPTER SIX•

# *Firefly*: So Pretty It Could Not Die

*Michael Marek*[*]

Joseph Campbell (1986) observed that creative artists are truly creative only to the extent that they are innovative. He added that we judge artists by both their technical innovations and their innovative insights. This postulate is nowhere more true than in episodic television, where there are few really new ideas. Rather, successful television results from creative re-invention of previous ideas and formulas. Indeed, I would suggest that in television as well as movie storytelling, it is innovative variation on existing formulas that should be the yardstick of creativity, regardless of how actual success is measured in terms of critical acclaim or profitability.

At the same time, the modern environment of home video means that the television and film stories of the past are no longer relegated only to the memories of those who once saw them. The studio vaults are open and a half-century of television (and almost a century of theatrical movies) remains fresh for discovery by new audiences. The stories are with us in the present and, in a way, are "mythology engines" in modern culture (Irvine & Beattie, 1998), creating many of our most recognizable metaphors and idioms. In this way, television and film contribute heavily to the way society defines and understands itself. Yet these mythology engines are rarely entirely new. Rather, they build on ideas and stories that have gone before. The *Veronica Mars* premise, for example, of a detective who is in high

[*] Dr. Michael Marek (M.A., University of South Dakota; Ed.D., University of South Dakota) is an Assistant Professor of Communication Arts at Wayne State College in Nebraska.

school is a variation on the "detective with a gimmick" formula that has been common on TV for decades. In the 1970s, *Cannon* was about a detective who was chubby and *Longstreet* was about a detective who was blind. In the 1980s, *Riptide* was about detectives who were best friends and *Simon & Simon* was about detectives who were brothers. More recently *Sue Thomas: F. B. Eye* was a police officer who was deaf. The multiple contemporary television police "procedural" dramas are also retellings of this same basic "gimmick" story premise. In the science fiction arena, Gene Roddenberry sold *Star Trek* to NBC as "*Wagon Train* to the stars" (Whitfield & Roddenberry, 1968, p. 23). Although the Old West images in *Star Trek* are subtle, they presented a creative common ground of understanding for the network, the studio, and the producers of the series. Through the process of retelling, the audience becomes familiar with the conventions and typical characters found in such dramas, but in order to create a successful new work based on a standard formula that has already taken on elements of mythology, producers must employ Campbell's strategy of creative re-invention.

The 2002 television series *Firefly* and its 2005 spin-off theatrical movie *Serenity* are prime examples of innovation (from here on when I say "*Firefly*," I refer to both, unless I indicate otherwise). They recombined familiar ideas in a way that feels fresh and attracted an instant cult audience. The *Firefly* characters and stories, as well as the visual and aural elements of the production, combine in an integrated whole, each component reinforcing the others. All of the elements, at one level or another, draw on mythic images that are alive and employed constantly in contemporary American culture. This innovation, and thus creativity, of *Firefly* stems from the way it surprises the audience with new twists on stock television situations.

The *Firefly* TV series premiered in the fall of 2002 on the FOX television network, but the network canceled it after only twelve episodes had been broadcast. The series is the story of the spaceship *Serenity*, a cargo ship of the *Firefly* class, centuries in the future. Her captain, Malcolm "Mal" Reynolds, fought on the losing side of an interplanetary war. Just as Confederate soldiers departed to the American West after the American Civil War, Mal retreated to the frontier of space. Surrounding Mal are his crew and passengers, each of whom is a finely crafted character. Several have mysterious backstories. The subsequent release of the DVD set of television episodes, including three not broadcast on FOX, sold so briskly that Universal purchased the movie rights from FOX and released a theatrical version of *Firefly* in 2005, with the entire original cast. The movie, titled *Serenity*, was critically acclaimed, beating box office hits such as *Batman Begins*, *The Chronicles of Narnia*, and *Harry Potter and the Goblet of Fire* to win a 2006 Hugo for "Best Dramatic Presentation, Long Form." The *Serenity* screenplay won the Nebula Award for the best script of 2005. Despite this recognition, the movie struggled to reach the "studio math" calculation of profitability. Although both *Firefly* and *Serenity* suffered from audiences below what Hollywood required for financial success (the TV series typically drew 2.5 to 3 million households), most of those people who did watch the original appearances of these works quickly became dedicated fans, referring to themselves as "Browncoats" for the cowboy-like coats worn as uniforms by the forces on Mal's side of the war. The reasons fans give include strong characters and stories. Even more so, fans cite the believable premise of *Firefly*, which I will show employed familiar mythic structures.

## The Ecology of Storytelling

In one activity in my "Writing for the Mass Media" class, I ask students to list as many elements as they can that add meaning in a dramatic TV program. A little brainstorming usually produces a lengthy list including the overall premise, acting, locations and set designs, camera angles and movement, "props," lighting, music, and sound effects. Each of these elements, done well, reinforces the plot as well as the allusions, metaphors, and symbolism of the story, thus strengthening the overall mythic qualities of the work. Some elements are subtle, functioning almost below the level of consciousness of the viewer, but some can be overt and powerful factors in the storytelling.

I employ a scientific metaphor to suggest that these disparate elements, when combined, make up an "ecology" of creative elements. Just as a biological ecology thrives or dies as the result of the interaction of its individual living things, a television program or movie succeeds or fails as the result of the successful combination of disparate elements into a unified whole. Most, if not all, of these elements have conventions associated with them, used repeatedly in Hollywood storytelling, such as story "formulas," techniques of acting, and standards for use of sound or lighting. From these multiple "stock elements," the movies and TV give us stories ranging from predictable cliché to inspired innovation. Each element plays a role in telling the story and should be considered as part of a unified whole.

I examine *Firefly* as a prime example of creativity by exploring four of the modern[1] myths it employs and the interaction of myth, character, and place in *Firefly*. Although I focus on the way *Firefly* employs mythology in its premise and character development, we should not forget that to effectively tell a mythic story in a visual medium like television or film, costuming, lighting, sound, set construction and decoration, and other technical elements must also reinforce the mythological underpinnings of the story.

## The Myths

Many scholars acknowledge that a new class of mythology has arisen in modern society. Gehmann (2003) observed that modern and traditional myths have different characteristics. Modern mythology masquerades as objective fact rather than imagination, providing society with a convincing construct of reality. Because they appear to reflect reality, these modern myths wield immense social power and may influence, support, reinforce, or challenge stereotypes and biases that exist in society (Abdullah, 2000). Leverette (2003) discussed modern mythology rising out of the media as a new kind of myth in which the producers of popular culture have replaced the ancient shaman in the role of purveyor of myth. Mass media now answer the questions addressed by traditional myths: Who are we? Why are we here? How should we live? The answers in the media, according to Leverette, are often driven by advertising and consumption rather than by philosophy. Yet dramatic storytelling in the visual media of television and film can also comment on, or serve as reflections for, what Campbell called "the possibilities of human experience and fulfillment in a given society at a given time" (Coustineau, 1990, p. 135).

This is the role played by *Firefly*, which reflects current society and culture through stories about working-class people in the future, traveling in spaceships to frontier settings. *Firefly* uses multiple structures, devices, and elements to reinforce four deeply socialized myths, each element contributing to a unified whole that produces a feel of authenticity for the series. Viewers find familiarity in these elements but are also surprised when scripts move at right angles to expectations derived from these socialized myths. The innovation of *Firefly's* premise stems from its strategy of surprising the audience with alterations of the stock situations, formulas, and even the clichés which viewers expect because of the pervasiveness of the myths in which they are rooted. In detail after detail, *Firefly* catches viewers off guard by going in different directions from what is expected. *Firefly* and *Serenity* make strong use of the four specific modern myths. Most important and obvious is the myth of the Old West, but *Firefly* also uses what I consider to be three additional modern myths—conspiracy theory, the myth of American hegemony, and the mythological structure of utopian versus dystopian visions of the future. *Firefly* often uses images, archetypes, and situations found in these four types of stories. Equally important, however, is the way viewers are surprised when the conventional use of these mythic themes is creatively challenged and usurped.

*Old West Mythology*. One of the most pervasive myths of America is that of the Old West, told in venues ranging from "dime novels" to big-budget Hollywood feature films, so much so that the Western is considered to be a genre. It arose from earlier cultural constructs about the wilderness and frontiers that also take on the role of myths. In these myths, the frontier contrasts with civilization and the city, a metaphor for modernity (Kushner, 1992). Fears of urbanization and modernity produced a romanticized view of rural life that found its purest form in an imagined frontier. *Robinson Crusoe*, first published in 1719, and *Swiss Family Robinson*, 1812, reflect this construct. In America, the myth of the frontier drove the "manifest destiny" of expansion across the continent, emerging in the eighteenth century, long before the Old West, and dominating popular culture throughout the nineteenth century (Pfitzer, 1995). Daniel Boone, a real person, was also the hero of fictional exploits in post-Revolutionary War America, making him an icon of early frontier America, when "the frontier" was barely west of the Appalachian Mountains. Details vary, but the formula of the frontier myth is that immigrants originating from an old or urban world conquer and domesticate the wilderness (Kushner, 1992). The heroes of the frontier stories often express their uneasiness with urbanization as the driving force behind their departure for the frontier. Even in Laura Ingalls Wilder's autobiographical *Little House* books, Pa Ingalls repeatedly moves his family farther west because there are "too many people" where they presently live (Wilder, 1935, p. 1), a true story expressing the frontier myth.

The Old West expression of the broader frontier myth first appeared as "dime novels" and in Owen Wister's serious 1902 novel *The Virginian*. The Old West is set roughly from the beginning of the Civil War in 1860 to the end of hostilities with American Indians in 1890. In these stories, the law is distant, and the protagonists function by their personal honor because law is an artifact of civilization that is not present on the frontier (Thornton, 2002). The Western, then, is usually a morality play with black-and-white values and has been called, in mythological terms, the Mount Olympus of American culture (Bentkowski, 1998). It is worth noting that although cowboy heroes often risk their lives to protect others, the term "cowboy" has also taken on certain negative qualities, suggesting one who is brash, unimpressed by the rule of law, and not completely under the control of society.

The Old West variation on the earlier frontier myth was embraced by the early twenthieth-century media (radio, television, and the movies), a fact that is central to its dominant place in society. The original radio version of *The Lone Ranger*, broadcast from 1933 to 1955, when radio was the dominant entertainment medium, fueled the imagination of a generation. The hero was "a tough hard man with a will of iron and an unswerving sense of justice" who had few weaknesses and subscribed to the highest ideals of American life (Dunning, 1976, p. 369). Gene Autry, Roy Rogers, and "Hopalong" Cassidy, among others, became staples of the Saturday matinee movie screens. Early television versions of popular radio Westerns, like *The Lone Ranger* and *Gunsmoke*, were supplemented by a host of series created for television, including *Maverick*, *The Rifleman*, *Bonanza*, and many others that are often little remembered a half-century later yet contributed to the deep impression the Old West myth made on American society.

Just as early twentieth-century storytellers reinvigorated the aging frontier myth, the advent of flight and space travel became the newest incarnation of the idea of the frontier. The language of the Old West was adapted by a new generation of journalists, entrepreneurs, and promoters to make the space age understandable to the public. Fictionalized accounts of travel in space found fertile storytelling ground in Old West imagery. *Firefly* is not the first portrayal of frontier-like conditions in science fiction settings. Robert A. Heinlein depicts agrarian planetary colonists farming and using nineteenth-century pioneer technology in novels such as *Farmer in the Sky, Time Enough for Love*, and *Friday*. Saturday matinee serials of the 1950s periodically showed cowboys fighting villains with fantastic origins, such as the twelve-part serial *The Phantom Empire*, in which cowboy Gene Autry battled foes from the scientifically advanced underground empire of Murania (Brower & Eason, 1935). More subtle Old West imagery also appears in such diverse non-Western movies as *Twister* and *Independence Day* (King, 1999).

*Firefly* draws heavily on the myth of the Old West, portraying a space-going frontier society in the wake of an interplanetary civil war, with dialog, costuming, sets, and other elements evoking the American West. As seen in *Firefly*, it makes sense that planetary colonists would have to get by on local resources, including what food they can grow or collect for themselves. The economies of interstellar trade are bound to result in a scarcity of supply ships. This gives the basic premise of *Firefly* a believable feel, reinforced by basing each character on an Old West archetype. Captain Malcolm Reynolds is the working-class cowboy, unimpressed by the laws of a distant government and idealistic in his worldview. Mal is cold and does not form close relationships, evoking a Clint Eastwood type of loner/drifter cowboy, his psyche defined by his loss of faith in people. His first mate, Zoë, is the cowboy hero's trail partner, assistant wagon master, or sheriff's deputy. *Serenity's* pilot and Zoë's husband, Wash, often serves as the comic relief character, evoking the cattle drive cook or the cowboy's "sidekick." The Companion, Inara, a high-class and socially accepted prostitute, is the saloon or dance hall girl. Crewmember Jane is the (male) outlaw mercenary. Simon and River, the brother and sister who are on the run from the government, are the Easterners or "city folk" who are out of their element on the frontier. Shepherd Book, the preacher with a secret past, represents a common Old West archetype. Kaylee, *Serenity's* "mechanic," could be a reincarnation of the Old West blacksmith. I discuss the characters in more detail later, but each is a fresh take on a standard archetype in Old West myth.

Another part of the Old West myth that is important to the *Firefly* storyline is a retelling of the American Civil War. When the real Civil War ended, many of the former soldiers went west to the frontier, becoming the cowboys and sometimes the outlaws of Old West myth. This theme of leaving civilization after a war is a fundamental element of Mal's background, expressed in the vocal theme of the TV series, evoking the conflict of war: "Burn the land and boil the sea, you can't take the sky from me."

It is worth mentioning that the closest real parallel to the spacecraft *Serenity* in the late nineteenth-century was a setting rarely seen in tellings of the Old West myth—the Missouri River "sternwheeler" riverboat. These craft, much smaller than the floating palaces of the Mississippi River featured in gambling yarns and *Huckleberry Finn*, were vital to the shipment of supplies to the gold fields of Montana from 1859 until the railroads superseded them (Petsche, 1974). One notable telling of the sternwheeler riverboat story, rarely remembered because it is not available on video, was the NBC television series *Riverboat* (1959–1961) starring Darrin McGavin as the captain with crewmember Burt Reynolds (Kotar & Gessler, n.d.). *Star Trek* fans invariably find it intriguing that McGavin's riverboat bore the name *Enterprise*. *Serenity* and the riverboat *Enterprise* each travel the frontier, carrying cargo and visiting pioneer settlements. Their crews each become involved with the people they meet along the way, often doing side jobs that are not cargo transportation. The foundations of the two TV series are similar tellings of the Old West myth.

Use of the Old West myth provides a familiar framework, both for viewers and for the writers and producers of *Firefly*. The comfort level produced by the archetypical characters and settings promotes the "willing suspension of disbelief" required for a successful science fiction story. But this deep familiarity also facilitates the *Firefly* storytelling style of surprising the viewer, such as the scene in the movie in which Mal is about to attempt a one-man rescue of Inara. If Mal were the stereotypical cowboy, he would tell Zoë, "If you don't hear from me, you take the ship and get out of here." Mal actually tells her, "...you take the ship and you come and rescue me," surprising and amusing the audience. In this way, the stories continually feel fresh and innovative.

While the myth of the Old West/frontier is the most powerful, other modern myths also inspire important elements of *Firefly*. By folding elements from other myths into the Old West themes, the narrative gains depth and richness that contribute to its authenticity. I examine three additional myths that add rhetorical meaning: the myth of American hegemony, conspiracy theory myth, and the myth of a utopian versus dystopian future.

*Myth of American Hegemony*. Historically, the myth of the frontier was a driving force behind the "manifest destiny" movement to occupy and control the North American continent. This movement was inspired by a Eurocentric understanding of North America as an empty wilderness waiting to be tamed by the superior Euro-American society (Ledford, 2004). Today, the United States remains ethnocentric, due in part to white racial identity (Gold, 2004), and to the country's postwar role as "leader of the free world" (Rajagopal, 2004). In television and the movies, this ethnocentrism resulted, for years, in casting primarily whites for lead roles, with a few minority actors relegated to supporting or "extra" parts. Hollywood is slowly learning the selling power of color-blind casting, with beer ads and situation comedies leading the way (Chong, 2006), but even when entertainment media do embrace minorities, the dominant framework tends to be black/white, marginalizing others, such as Asian Americans (Chang, 2006). Culture in the United States has internalized a powerful perception of American hegemony in political, social, cultural, and technological realms (Pahre, 2005).

In science fiction about space travel, particularly as seen in America on the screen, the modern hegemonic myth is that the heroes who explore space are usually Americans, or descendants of the American heritage. In *Forbidden Planet*, *2001: A Space Odyssey*, *Star Wars*, and a host of less prominent movies about space travel, the main characters are white with exemplary Midwest American accents. Characters of British or other nationality sometimes appear, such as Patrick Stewart's Jean-Luc Picard, Alec Guinness' Obi-Wan Kenobi, or the Russian characters in *2001: A Space Odyssey* and *2010: The Year We Make Contact*, but they have historically been the exception to the rule. In part, this trend results from the purely business practice of producing entertainment for a mass (majority) audience and thus featuring characters to whom the majority audience relates, yet it also reflects the American public's deep conceptualization that the actual American space program is the "best" among those of the various countries exploring space and the one most likely to produce future accomplishments. This understanding about the primacy of the United States reflects American hegemony, functioning the way Gehmann (2003) defines modern myth, because it appears to reflect reality and represents a cultural understanding. It is not, however, unique to the United States. The great superpowers of history dating back to the Roman Empire, and maybe beyond, all held the same conceptualization that their way was the best.

The central proposition of *Firefly*, however, counters this hegemonic myth, thus repeatedly surprising viewers. In *Firefly*, the American/Western culture has blended with the Chinese/Asian culture, as the two great political powers explored space. This idea plays on China's first manned space flights in the early twenty-first-century, as well as with the growing Western understanding of the power of the Chinese economy. As a result, characters in *Firefly* frequently voice brief Mandarin phrases, often as a substitute for profanity not allowed in English in the highly regulated American television media. The assumption that even working-class people of the future will be fluently bilingual makes sense, but each time Chinese dialogue appears in *Firefly*, it is an opportunity for the audience to realize that they are watching an innovative view of the future.

This strong Chinese flavor of *Firefly* contrasts with the relative absence of Asian culture from American television (Mathy, 2001) and with the assumption of most science fiction that space travel in the future will evolve out of the American space program and be dominated by Western influences and ethnic types, itself a product of the perception of American hegemony. The assumption of American dominance of the future of spaceflight is deep in the psyche of American science fiction fans, rooted in the broader social construct in the United States that the American government, institutions, and culture are "the best."

With regard to the myth of American hegemony, *Firefly* provides a counter-hegemonic critique. The cult following developed by *Firefly* suggests that many fans are willing to embrace the alternative suggestion that Chinese culture will become more prominent in the future affairs of the human species. Every appearance of a Chinese influence in *Firefly* gives the viewer pause and reinforces the impression of innovation.

*Conspiracy Theory Mythology*. Another key modern mythology drawn on by *Firefly* is that known as "conspiracy theory," the belief that powerful and malevolent elite groups secretly operate in society. This idea pervades American cultural understanding of societal problems, including the JFK assassination, high gas prices, UFOs, and a host of other negative events in which the alleged conspirators are mysterious, powerful, and never seen (Nelson, 2003), and it fits

Gehmann's (2003) definition that modern myth masquerades as objective fact, providing a convincing construct of reality. I see the broad idea of conspiracy theory as a mythic structure, with each individual "case" of conspiracy serving as a "retelling" of that foundational myth. The simple explanation of a conspiracy is often more gratifying than the complexity of learning about the case at the level of detail required to understand the nuances of the truth. Elements in a society may also employ the conspiracy myth for political purposes, using myth and symbol as a means of constructing an enemy or challenge to overcome. By rhetorically defining an enemy in simple terms, often through demonization, a society can achieve unity because, without a clear external enemy, the society may focus disruptively on internal political enemies (Brustein, 1991). Regimes throughout history have used this technique (having the practical effect of connecting conspiracy theory to the influences of hegemony), allowing their people to see themselves as the most righteous when their villains can be defined as the most nefarious. The World War II Allied propaganda effort, the "domino theory" used to justify the Vietnam War, and the current Bush administration's efforts to connect the 9/11 attacks with Saddam Hussein can all be seen (depending on your politics) as using this technique of rhetorically defining an enemy in simple, negative terms.

Conspiracy theories have three central characteristics (Steiner & Edmunds, 1979): (1) they have a surface conformity to real life events; (2) they are easy to prove via distorted (nonlogical) deduction; and, (3) they are based on primitive understanding of the dynamics of large organizations. Conspiracy theorists see large complex systems in terms of simple small-group behavior. Because conspiracies are attractive explanations for the wrongs of society, more and more entertainment enterprises are cashing in on the willingness of people to embrace conspiracy theory (Birchall, 2001), with conspiracy storylines populating many contemporary TV series and movies.

The fundamental mystery of *Firefly*, revealed in the feature film *Serenity*, is that the central government sent millions of colonists to a new planet, Miranda, and put chemicals in their food and air intending to calm the population and to weed out aggression, making them peaceful and happy. Instead, most of the colonists died, while a few became the ultra-aggressive "Reavers," wantonly raping and killing wherever their spaceships found victims. The seventeen-year-old River Tam, in a secret government academy for psychics, has read the minds of visiting members of parliament responsible for the secret Miranda disaster. When her older brother, Simon, rescues her from the academy, River has developed an extreme psychosis and does not realize what information she carries. Hiding out on *Serenity*, Simon and River are the subjects of repeated attempts by government operatives to recapture them.

The Blue Sun Corporation, referred to only in background visual elements not central to the story, is an undeveloped element of the conspiracy in *Firefly*. It is apparently intended to be an interplanetary mega-corporation with its fingers in every element of daily life. The implication is that Blue Sun has ties to the Alliance, and possibly to the disaster on Miranda, although this is not stated overtly. The inclusion of a corporate element in the *Firefly* premise makes use of the recurring presence of major corporations in the mythology of conspiracy theory. The conspiracy that underlies the TV series is as shadowy and distant as any, unfolding incrementally, episode by episode. In producing the movie, storytelling required that the conspiracy unfold over the course of two hours. In this way, movie patrons who had not seen the TV series would still understand and benefit from the full evolution of the conspiracy storyline. The movie presents a single representative of the conspiracy as the antagonist, the villain, the opponent of the *Serenity* crew. Known only as The Operative, this character serves the goals of the conspiracy with ruthless methods but provides a dramatic focus for the viewer.

Perfectly in keeping with Gehmann's (2003) definition, *Firefly* presents shadowy people of power, always acting off-stage and in secret, but having a profound impact on the lives of citizens. But in *Firefly's* use of conspiracy elements, the Miranda conspiracy stands apart from the stock "bad guys secretly doing bad things" found in most incarnations of the conspiracy myth. Rather, the innovation, and therefore the creativity of its use in *Firefly*, is in the motivation of the conspirators, as presented in the intrinsic motivation of The Operative, which is to create "a better world." Consider the following dialog between Mal and The Operative after several groups of people who would have given the *Serenity* crew refuge are killed:

> Mal: I don't murder children.
> Operative: I do...if I have to.
> Mal: Why? Do you even know why they sent you?
> Operative: It's not my place to ask. I believe in something greater than myself. A better world. A world without sin.

Mal: So me and mine gotta lay down and die so you can live in your better world?

Operative: I'm not going to live there. There's no place for me there, any more than there is for you. Malcolm, I'm a monster. What I do is evil. I have no illusions about it, but it must be done.

The Operative's "end justifies the means" ethics, enunciated here, holds that the greater good of society allows comparatively small moral transgressions. He believes so strongly in this greater good that he does not question the methods. At the end of the movie, when The Operative learns the truth of Miranda, he pragmatically abandons the parliamentary conspirators with a "no hard feelings" attitude toward the people he had until recently been persecuting.

So, while conspiracies usually are understood to be bad, harmful, and malevolent, we find that the *Firefly/Serenity* conspiracy (at least as understood by The Operative) is paradoxically motivated by ending violence and evil. This "twist" on the conventional conspiracy helps make The Operative a strong character but is also an innovative, thus creative, "take" on typical conspiracy mythology.

*The Utopian versus Dystopian Myths of the Future.* The final modern myth that plays a role in *Firefly* is a two-sided coin—what I call the Utopian versus Dystopian visions of the future that have been common in society at least since the biblical *Book of Revelation* (Mourby, 2004), and are common in science fiction. I suggest that utopia and dystopia function in society as a myth because, together, they meet the criteria of Gehmann (2003) by masquerading as objective fact rather than imagination and of Abdullah (2000) by appearing to reflect reality, and thus wielding immense social power. Regardless of whether you believe, for example, that global warming is true, it is a dystopian view of the future and is slowly becoming a juggernaut in the arena of American public policy. Translated into the realm of fiction, utopian and dystopian stories about the future tend to portray one of two extremes. Either the future is dark and post-apocalyptic or it is positive and desirable. The utopian vision of the future, in fiction and elsewhere, is constructed out of our hopes, wishes, and intentions (Ollman, 2005). The dystopian view shows a future that is disturbing and populated with a "human society that is truthful to the vagaries and vicissitudes of human culpability, failure, and compromise" (Wright, 2004, p. 233).

Stories in utopian settings (or dystopian settings, for that matter) are so common on television, in the movies, and in popular print fiction that we automatically anticipate the elements of such a story. Utopian stories tend to depict high-tech future worlds and advanced societies with many of today's problems resolved. The epitome of this utopian vision appears in *Star Trek*, in which money is no longer used, apparently because energy is so easily available that anyone can easily fabricate or "replicate" anything they might want. *Trek* fans have built up a highly detailed understanding of the *Star Trek* universe that they use to evaluate the quality of new Trek episodes or stories (Marek, 2004). Dystopian science fiction abounds on the large and small screen, including *Battlestar: Galactica* (2004); *The Matrix* trilogy; *I, Robot*; and *Blade Runner*. Yet dystopia and utopia are symbolically linked, just as the dystopic world vision of Mr. Hyde contrasts with utopia's Dr. Jekyll (Mourby, 2004). Dystopian stories serve as warnings and often contain implicit elements of hope, as characters struggle to solve or escape the dystopian problems.

In *Firefly*, we see a tension between utopian and dystopian elements of oppression versus freedom. Society has divided into haves and have-nots. In the central core worlds of the Alliance, we see a utopian reality of affluence. In the outer worlds, poverty is widespread; slavery and indentured servitude are common, and crime and violence are rampant. Most of the worlds visited by the *Serenity* crew are primitive and depressing. Most of the *Serenity* crew seek a positive, utopian, future. Constant confrontations arise, however, from the reality of existence in the dystopian environment of the outer planets.

In the *Firefly* television series, these confrontations are episodic in nature, mirroring the nature of television. The *Serenity* crew meets crime lords, unscrupulous power brokers, members of the working class who have little hope, and the "boogie men" Reavers as the series proceeds week to week. It is not until the movie, however, that the "dots are connected" and we understand the true nature of the dystopian conspiracy that has killed millions of people in secret. In the movie, Mal is the agent of utopia, reaching the point where he idealistically takes a stand against the dystopian parliament, personified by The Operative. By contrasting the utopian and dystopian struggle for the future, *Firefly* presents a story that has an authentic feel, plus the opportunity to offer social commentary.

## The Role of Character and Place in Modern Myth

The phenomena of myth and "place" are closely linked together in a given culture (Borsay, 2006), and particularly in the fiction of a culture where myth and place are key ingredients that writers use to construct characters. In the modern myths produced by television and the movies, main characters almost always have places which they inhabit or from which they originate and that serve to define them. These places make statements about the role of the character in the myth. The Cartwright family of *Bonanza*, an Old West myth, is defined by the Ponderosa ranch and particularly by the aristocratic log lodge that serves as their home. In the conspiracy myth *The X-Files*, Fox Mulder is defined by his basement office and his solitary apartment, underscoring his isolation. The heroes of the spaceflight myth *Star Trek* are tightly tied to the starship *Enterprise* as the place that defines their society and background. In *Firefly*, the spacecraft *Serenity* also serves as the unifying place for the characters, where they find common ground and peruse their mutual relationships. Each, however, has a unique place that defines the character psychologically. These places are based in mythic archetypes, particularly rooted in the Old West/frontier myth, but also in the other three central myths on which *Firefly* is based.

Mal Reynolds is a complex character based on the Civil War veterans who went west to the frontier, a common character in the Old West myth. Emotionally, Mal has never left Serenity Valley. Not unlike the real prisoner-of-war camp Andersonville, Serenity Valley is the disastrous site of the final collapse of the Independent rebellion, of the deaths of most of Mal's subordinates, and of his loss of faith. The place defines him as a character. The consequences of what happened there are so burned into his psyche that he has never been able to move past them. Even his decision to name his ship *Serenity* demonstrates that he is still there, emotionally hiding in a foxhole.

River is a victim of the conspiracy, defined by the Alliance Academy and what happened to her there, her brain surgically altered to make her a clairvoyant assassin. Every scene involving River, such as her violent outbursts and her use of psychic abilities, stems from what the conspiracy did to her in the academy. River also experiences place with more richness than the other characters, almost at a spiritual level. She often walks barefoot, giving her tactile contact with her immediate surroundings. She moves slowly and gracefully, like a dancer, as she moves through spaces. She is a product of the utopian side of the *Firefly* future, transformed by the conspiracy, and is living in the dystopian side.

Both Simon and Inara are defined by the utopian places they left behind. Simon was an elite surgeon in a central planet hospital. He would still be there had he not needed to run from the Alliance to save River from the conspiracy. He is out of his element everywhere else, except when performing medical services. Inara, meanwhile, was rapidly advancing in the hierarchy of Companions on a central Alliance planet but left abruptly for reasons never explained on screen. Her clothing, the decor of her quarters, and her personal attitudes always subtly seem to be an effort to recreate her former utopian environment.

Zoë has two defining psychological places, one professional and one personal. In keeping with the Old West mythology, she is the "partner" to the hero, the trail buddy, the faithful subordinate. As such, one of her places is at Mal's side, similar to the way that in *Star Trek* Spock's perfect place was at Kirk's side, as observed by Edith Keeler. Not just a sidekick, Zoë is called a "warrior woman" by her husband—strong, cool in a fight, and capable with a variety of weapons, a product of the dystopian side of the *Firefly* future. But Zoë is also a wife with dreams of motherhood, an emotional place that can be at odds with her profession. The fifteen episodes of *Firefly* did not explore this balance in detail, but had the series continued, it could have become a metaphor for the challenges faced by professional women in real life.

Wash, on the other hand, is primarily defined by his place at *Serenity's* controls. He is the consummate pilot, with an almost Zen-like ability to make the ship an extension of himself. But his place behind the controls is also very localized and personal. His toy dinosaurs, for example, rest around the control panel, almost becoming part of the equipment. When he dies in the movie, the dinosaurs remain, as if it occurs to no one to remove them. Other than during the intense piloting scenes, Wash is the "everyman" of the story. His signature Hawaiian shirts and his humor underscore the American side of the dual American/Chinese culture.

At face value, Kaylee's place is the engine room, because of her intuitive grasp of equipment. But even more so, her psychological place is the rural area where she grew up. She is based on the All-American country girl, never having ventured far from the farm before joining the *Serenity* crew. She is far from innocent but naturally sees the good in every person and situation, making her the emotional heart of the *Serenity* crew. More than anyone else, she sees through Mal's chilly dystopian exterior, even though he tells her, without conviction, "I'm a mean old man." Kaylee is therefore also a principal voice of utopian cultural values, despite her nontraditional "grease monkey" role on the crew.

If Kaylee is the heart of the crew, Shepherd Book is the soul, the most spiritual of the characters. He has left Southtown Abbey because he has "been out of the world for a spell; like to walk it a while." In the *Firefly* pilot, he seems remarkably unsophisticated in the ways of the world, but it slowly develops that he has a secret past, never revealed, that is significantly darker. Was he an Alliance military officer? An Operative, himself? What led to his spiritual conversion? This unrevealed place, and the emotional journey that led him to leave it, defines Book's character, making him also based in Old West myth, which is replete with characters seeking redemption from past sins, including former outlaws who are "going straight."

Jane is the only member of the crew who seems to have no defining place in his past. Because he is deeply self-centered and lives in the here and now, one might say that his place is wherever he is in the present, or that his place is inside himself. He is the mercenary, changing allegiances as convenience and profit dictate, and a product of the dystopian side of the *Firefly* future. Had the TV series continued, he would have gone through a change in mindset, as he does in the movie. When Mal announces his plan to cross the Reaver and Alliance lines to deliver word of the Miranda disaster to the entire system, Jane gives his unqualified support, even though there is no profit for him. Like Book, Jane is beginning to redeem his past dishonorable ways, evoking Old West cowboy villains turned hero.

The ship itself is also a place that is important to *Firefly* as a storytelling element in individual episodes. One of the most complex of the *Firefly* TV episodes is "Out of Gas." Environmental systems on *Serenity* are damaged and the crew expects to die of suffocation if they remain onboard. The story begins in the middle, intercutting flashbacks showing how the situation arose, as well as flashbacks to more distant times showing how several of the characters first joined the *Serenity* crew. The episode is about claustrophobia, about people trapped in a place in which they will eventually die. There is also a pivotal scene in which Mal refuses to leave *Serenity* as the others prepare to depart in shuttles. The rich visual elements repeatedly emphasize Mal moving through the spaces of the ship, including carefully closing every door behind him as he heads to the bridge after the others have departed. The story allows each crewmember to express thoughts that resonate with his or her actual and emotional places of origin, thus evoking the modern myths that inform their characters.

The final produced television episode employs place in even more direct terms. "Objects in Space" is probably the most philosophical episode of the TV series. It tells the story of a psychologically unbalanced bounty hunter, Jubal Early (named after a real Civil War general), who boards *Serenity* in deep space to capture River, an allusion to Old West bounty hunter stories. Early, who may also be mildly psychic, has traits in common with River. He experiences place directly, even touching a catwalk railing with his tongue to connect with it. He asks whether River's room is still her room when she is not in it, whether the function of a place really defines it in an absolute sense. After Early incapacitates most of the other crew members, River claims to have merged with *Serenity* and literally "becomes the ship." It develops that she is verbally fencing with Early and she has actually left *Serenity* and taken control of Early's nearby craft. Symbolically, however, River's claim to have become the ship actually means that she has taken the leadership of the *Serenity* crew, because, to her, the ship is the people. She directs Kaylee to escape her bonds and release the rest of the crew. She directs Mal to position himself in order to defeat Early.

The message about place in "Objects in Space," and in *Firefly* as a body of work, is that our understanding of place is a social construction of the people who live in, use, or abide in the place. This understanding is important because place informs all of the modern myths employed by *Firefly*—the places of the Old West, the places that depict a bright or dark future, the places that reveal or hide the conspiracy, and America itself as a source of hegemony.

## Retelling the Myth

TV and movie fiction, at its best, has an elemental formula. It provides stories about nuanced characters meeting and overcoming challenges, in the process finding that something has changed within them, usually for the better. This formula fits perfectly with Campbell's definition that mythic narratives are "metaphorical of the possibilities of human experience and fulfillment" (Coustineau, 1990, p. 135). The exploration of human experience and fulfillment in fiction is rooted in the humanity of the characters, meaning that for a TV series or movie to be successful, we must come to care about the characters and see them transformed.

My examination of *Firefly* has shown that two fundamental sources of mythic images are employed in television and movie storytelling. The first source is deeply socialized myths of modern society, such as the four modern myths employed in *Firefly*, which can be sources on which characters and storylines may be based. The second source in which visual storytelling may find inspiration is in the retelling and reimagining of other television and movie productions, which through retelling take on an aspect of myth themselves. However, because creativity is rooted in innovation, the best visual storytelling is that which does not simply make superficial changes in previous incarnations of the myth but, rather, leads the viewer in unexpected directions.

*Firefly* the TV series is a successful application of this elemental formula, with a complex set of characters and a monumental challenge, slowly unfolding. In the movie retelling of the story, the same characters and challenge exist, compressed to make the character development all the more obvious, and changing each of them, the captain most of all. The fact that a cult audience of at least a couple of million people developed for *Firefly/Serenity* suggests that the producers were successful in their creation of mythology for the new millennium, profitability considerations notwithstanding.

Although I have primarily analyzed *Firefly* in terms of how it employs mythic images that are socialized in contemporary culture, there is another way to apply the concept of myth to this TV series and related movie—that they have, themselves, become myths to their fans, stories worth retelling that have messages relevant to modern culture. Classical myths have come down to us from multiple sources, in part because they are so often alluded to by poets, essayists, and orators (Bulfinch, 1978), meaning that the act of retelling is central to the production and continued relevance of mythology. Similarly, a modern story told only once does not function as myth. It is its repetition, embellishment, and sometimes its reinterpretation that causes the story to become mythic.

Although production on *Firefly* ended in 2002 (and it is unlikely that a sequel movie will ever be made) the myth that is *Firefly* continues to be retold. In May 2007, the compiled TV episode boxed set remained on Amazon.com's top-100 best-selling DVD list. *Firefly* comic books sold well in 2006, as did CD soundtracks of the television series and movie. Several references from *Firefly* have been adopted by fans, and among the 2.5–3 million member fan base, at least, are instantly recognizable and applicable to other situations. When production of *Serenity* was announced, fans immediately began calling it the "BDM," short for "Big Damn Movie," an allusion to Zoë's claim to be "Big Damn Heroes" in the episode *Safe*. Use in several episodes of "Shiny" as slang for anything good or positive is also immediately recognized by fans.

The best way to understand the movie *Serenity* is as a retelling of the story of the TV series, not simply a continuation, and therefore as a retelling of the myth. Yes, it features the same actors and has no inconsistencies with the TV series, but it compresses into two hours the overarching story and character development that would have unfolded over several seasons had the TV series continued. It is not a coincidence that the name of the movie is also the name of the two-hour pilot episode of the TV series, because it returns to the beginning as its starting place for the new telling.

## Conclusion

I began by suggesting that in movie and television storytelling, it is innovative variation on existing formulas that should be the yardstick for creativity, understanding that creativity and commercial success are not always the same thing. My examination of *Firefly* suggests this assertion can be revised to say that it is innovative variation on established societal myths that measures creativity in television and film. These myths may be the deeply socialized myths observed by many scholars, such as the Old West, or they may be new takes on the standard formulas of episodic television, which in turn can establish their own functional myths. So, while myth is the anvil on which innovation is forged, innovation, in turn, forges new myth, as the most compelling creations take on lives of their own, and through retelling become socialized modern myths in their own right.

## Note

1 In social and historical terms, the "modern era" is generally held to have begun with the end of the medieval period, the Protestant Reformation in Europe, or the Industrial Revolution, and to have lasted to the latter half of the twentieth century, if not to today (Collins, 1992; Trainor, 1998). As I use the term "modern myth" in this chapter, it is not so much about the date range of the modern period but rather is about a way of thinking of society. In this context, to say that a myth is "modern" does not limit it to recent years. It may have roots that date back 100–200 years ago, yet still function as a myth today. On the other hand, where I use the term "contemporary" in this chapter, I am referring to a time period restricted to the past several years.

## References

Abdullah, M. (2000). Media literacy. *ERIC Digest D152*.

Bentkowski, T. (1998). Lone ranger. *Life, 21* (2), 25–27.

Birchall, C. (2001). Conspiracy theories and academic discourses: The necessary possibility of popular (over)interpretation. *Continuum: Journal of Media & Cultural Studies, 15* (1), 67–76.

Borsay, P. (2006). New approaches to social history, myth, memory and place: Monmouth and Bath 1750–1900. *Journal of Social History, 39* (3), 867–889.

Brower, O. & Eason, B. R. (Directors). (1935). *The phantom empire* [Motion Picture]. United States: Mascot.

Brustein, R. (1991). The use and abuse of multiculturalism. *New Republic, 205* (12/13), 31–34.

Bulfinch, T. (1978). *Bulfinch's mythology*. New York: Avenel Books.

Campbell, J. (1986). *The inner reaches of outer space*. New York: HarperCollins.

Chang, J., (2006). Diversity is NOT black and white. *Chinese American Forum, 21* (3), 37.

Chong, K. (2006). Look, over there. It's a black guy! *Maclean's, 119* (27/28), 90.

Collins, R. (1992). The rise and fall of Modernism in politics and religion. *Acta Sociologica, 35* (3), 171–186.

Coustineau, P. (Ed.). (1990). *The hero's journey: The world of Joseph Campbell*. San Francisco: Harper & Row.

Dunning, J., (1976). *Tune in yesterday*. Englewood Cliffs, NJ: Prentice-Hall.

Gehmann, U. (2003). Modern myths. *Culture and Organization, 9* (2), 105–119.

Gold, S. (2004). From Jim Crow to racial hegemony: Evolving explanations of racial hierarchy. *Ethnic & Racial Studies, 27* (6), 951–968.

Irvine, S. & Beattie, N. (1998). Conspiracy theory, pre-millennium tension and the X-Files: Power and belief in the 1990s. *Social Alternatives, 17* (4), 31–34.

King, G. (1999). Spectacular narratives: *Twister, Independence Day*, and frontier mythology in modern Hollywood. *Journal of American Culture, 22* (1), 25–39.

Kotar, S. L. & Gessler, J. E. (n.d.). Riverboat. Retrieved June 1, 2006, from http://www.darrenmcgavin.net/riverboat.htm

Kushner, H. I. (1992). The persistence of the "frontier thesis" in America: Gender, myth and self-destruction. *Canadian Review of American Studies, 23* (1), 53–83.

Ledford, K. E. (2004). "Singularly placed in scenes so cultivated": The frontier, the myth of westward progress, and a backwoods in the mountain south. *American Transcendental Quarterly, 18* (3), 205–222.

Leverette, M. (2003). Midriffs, mooks, and myth: Re-imagining Campbell's hero in the post-MTV era. *Conference papers of the International Communication Association 2003 Annual Meeting*, San Diego, CA.

Marek, M. W. (2004). *Enterprise is the* "New Coke" *of* Star Trek. Posted April 10, 2004 at The Great Link Web site: http://www.greatlink.org/springbreak.asp

Mathy, J. P. (2001). The popularity of American culture. *Sites: Journal of the Twentieth-Century/Contemporary French Studies, 1* (1), 141–156.

Mourby, A. (2004). Dystopia: Who needs it? *History Today, 53* (12), 16–17.

Nelson, J. S. (2003). Conspiracy as a Hollywood trope for system. *Political Communication, 20*, 499–503.

Ollman, B. (2005). The utopian vision of the future (then and now): A Marxist critique. *Monthly Review: An Independent Socialist Magazine, 57* (3), 78–102.

Pahre, R. (2005). Hegemony and the international economy. *Comparative Sociology, 4* (3–4), 451–477.

Petsche, J. E. (1974). *The steamboat Bertrand: History, excavation and architecture.* Washington: National Park Service.

Pfitzer, G. M. (1995). The only good alien is a dead alien: Science fiction and the metaphysics of Indian-hating on the high frontier. *Journal of American Culture, 18* (1), 51–68.

Rajagopal, A. (2004). America and its others. *Interventions: The International Journal of Postcolonial Studies, 6* (3), 317–329.

Steiner, J. F. & Edmunds, S. W. (1979). Ascientific beliefs about large organizations and adaptation to change. *Academy of Management Review, 4* (1), 107–112.

Thornton, B. (2002). The cowboy myth. *American Spectator*, 35 (3), 20.

Trainor, B. (1998). The origin and end of modernity. *Journal of Applied Philosophy, 15* (2), 133–144.

Whitfield, S. E. & Roddenberry, E. (1968). *The making of* Star Trek. New York: Ballantine.

Wilder, L. I. (1935). *Little house on the prairie*. New York: Harper & Row.

Wright, A. (2004). An ambiguous utopia. *Political Theology, 5* (2), 231–238.

•CHAPTER SEVEN•

# "Small World": Alex Proyas' *Dark City* and Omnitopia

*Andrew Wood*[*]

A woman encounters a stranger at a dock on the river's edge. Looming over them, framing them, the city glows at night. The woman looks over the water; she is fearful that her husband is lost and may not return. The stranger knows this and much more. He possesses a secret knowledge that both the woman and her husband do not know each other at all. Bound up in faked photographs and implanted memories, their love is an experiment concocted for an alien purpose. The stranger understands the experiment because he is one of the researchers. To the woman, the city is a labyrinth that has caught her husband. To the stranger, the city is a laboratory in which people and places are warped, edited, and erased, a place where memory and identity constitute a mutable construct. The stranger visits the woman and toys with her, speaking the words implanted in her husband:

> Mr. Hand: There used to be a ferry when I was a boy. Biggest thing you ever saw, lit up like a floating birthday cake.
> Emma Murdoch: That's just what my husband once said to me, on this very spot.
> Mr. Hand: Where is your husband now?
> Emma Murdoch: I wish I knew. What brings you here?
> Mr. Hand: I met my wife at this place.
> Emma Murdoch: It's where I first met my husband.
> Mr. Hand: Small world.

[*] Dr. Andrew Wood (M.A., Ohio University; Ph.D., Ohio University) is an Associate Professor of Communication Studies at San Jose State University.

The world is smaller than he admits. Its limits and purposes stretch no further than the bounds of the city that encloses them both.

This is a scene from *Dark City*, a 1998 film by Alex Proyas that joins Fritz Lang's 1927 *Metropolis* and Ridley Scott's 1982 *Blade Runner* as an indispensable text for students of urbanity in popular culture.[1] With its potent visual iconography and illustration of contemporary trends, *Dark City* offers a cinematic critique of enclavic urbanity that demands closer attention than it has thus far received. Indeed, like its major characters, *Dark City* initially encountered a closed system, a small number of viewers who were drawn to its peculiar convergence of noir detective procedural, psychological thriller, and science fiction effects-fest. The film's plot focuses on an amnesiac accused of murder and his efforts to learn his identity. As he seeks clues about his past, the protagonist encounters a group of "strangers" who possess a startling secret. His struggle with the strangers reveals truths about his character and his world that inspire the audience to question the nature of reality itself. This presumes, of course, that an audience would sit through the film.

A box office flop in the United States, *Dark City* inspired unflattering comparisons to *The Matrix*, despite appearing on screens one year prior to that hit film.[2] However, mirroring the spiraled evolution of *Dark City*'s narrative, the film managed to grow into a cult favorite, aided by rentals, midnight screenings, and film critic Roger Ebert's personal evangelism.[3] Paralleling its popular growth, *Dark City* has also begun to attract international scholarly attention from researchers in American studies (Blackmore, 2004; Gerlach & Hamilton, 2004), cultural studies (Milner, 2004), and media studies (Hayles & Gessler, 2004; Higley 2001; Tripp, 2005), along with English and linguistics (Marsen, 2004). But more work remains to be done, particularly on the film's unique statement about the changing myth of the frontier.

In this chapter, I seek to contribute to this expanding conversation by focusing my attention on *Dark City*'s comment on the changing nature of urbanity. This analysis grows from a larger interest in the role of myth in cinema, the means through which broad ideas of how one should live become encapsulated in iconic figures, tropes, and types—in this case, the frontier and the city. Here, I argue that *Dark City* provides a lens to view contemporary urbanity with its illustration of an enclosure whose inhabitants reside in a kind of daze, manipulated by forces they cannot see. Initially, this analysis requires a brief overview of the mythical frontier as a counterpoint to the urban maze portrayed by *Dark City*. Following this overview, I employ an omnitopian framework to map out the means though which the frontier has been banished by the cinematic city, leaving in its place an enclave from which one can scarcely hope to escape or even to imagine the possibility of departure. I conclude with remarks on how *Dark City*'s critique of omnitopia, while suggesting a return to the frontier, rests on a troubling implication for heroes who cannot be termed human.

## Birth of the Mythical Frontier

Scholarship on the frontier myth almost invariably begins with the "frontier thesis" posited by Frederick Jackson Turner at a meeting of the American Historical Association held at Chicago's 1893 Columbian Exposition. Prior to Turner's presentation, the vast majority of historians seeking to uncover the development of the American mind focused their attentions to the nation's European roots. The prevailing wisdom held that America's national identity sprang from its association with the Old World and grew always in relation to that parentage. Turner, in contrast, proposed that the frontier rather than the homeland offers more insight into the New World.[4] Turner differentiated the European and American notions of frontier: the former referred to a heavily guarded boundary; the latter indicated the line demarcating freedom from convention. Rather than separating two established entities, the American frontier opened up possibilities beyond what had been established. Consequently, those who pursued the frontier did not seek to protect the conventional order but rather faced adversities and overcame challenges in a manner best illustrated by the persona of the frontiersman. While often quoted, this passage by Turner offers apt illustration:

> That coarseness and strength combined with acuteness and inquisitiveness; that practical, inventive turn of mind, quick to find expedients; that masterful grasp of material things, lacking in the artistic but powerful to effect great ends; that restless, nervous energy; that dominant individualism, working for good and for evil, and withal that buoyancy and exuberance which comes with freedom—these are traits of the frontier. (p. 37)

Turner's thesis concludes by announcing the *closing* of the frontier due to the nation's rapid urbanization, further differentiating urban and unsettled America.[5]

To Turner's nineteenth-century listeners and readers, the bright and open spaces of the frontier suggested an immediate contrast to the dominant myths of urbanity that they held. To the Victorian mind, the urban core was truly a "dark city," a container of dangers and terrors that threatened the souls of its inhabitants. Narrow streets, unlit alleys, violent crime, and anonymous "strangers" resulted in chaotic moments and social confusions. Whereas in the mythical countryside one knew one's neighbors and, by extension, one's "place," the city exploded all norms. Each footfall along its arcades and busy boulevards resulted in chance encounters with no guiding narrative or consistent patterns to follow. Brimming with immigrants and bursting with chock-a-block growth, the city was a maze, a labyrinth, a trap. As a physical and rhetorical site of contrast, the frontier revealed the possibility to escape the city and seek personal transformation.

Within the field of communication studies, Turner's students and advocates, along with historians of his legacy, eventually carved out their own place to explore the impact of the frontier myth upon American oratory and media. In his 1941 essay "The Speech of the Frontier," Edward Everett Dale announced the need for communication scholars to study the impact of the sometimes salty but frequently shrewd frontier discourse on contemporary speech.[6] A year later, Robert D. Clark (1942) offered a somewhat more sophisticated reminder of the impression of frontier talk on the great political speakers in American history.[7] But it was Ronald H. Carpenter who most convincingly and thoroughly explored the frontier myth from the standpoint of communication scholarship. Carpenter (1977) examines that rare phenomenon of a piece of scholarship by a seemingly unknown historian inspiring national discussion and altering the common conception of the American identity.[8] Inspired by his contribution while departing from Carpenter's historigraphic emphasis, Janice Hocker Rushing (1983, 1986, 1989) launched a mini-revival of communication scholarship on the frontier in the 1980s with articles that explored mediated manifestations of the frontier myth. Importantly, she inspired the transition of attention from the Old Frontier (the "West") to the New Frontier (space).

Of particular interest to this chapter, Rushing offers insight into efforts to transplant Western themes into urban settings, as epitomized by the 1980 film *Urban Cowboy*. She notes that this type of film attempts to "modernize" frontier life for contemporary audiences, "[b]ut urban life lacks space and thrives on industry; both conditions are anathema to the wandering soul of the West, free of technology and at one with nature" (p. 266). Here, Rushing emphasizes that the frontier, whether old or new, rests on a dichotomy between desert and town (and, within the town, the brothel and the schoolhouse). In this manner, the frontier demands a counterpoint—the home, the city, the hometown—for its definition. Conversely, urbanity requires the frontier—the desert, the ocean, the outback—as its own mythical counterpoint. Yet, as I elaborate below, contemporary urbanity seems to have rid itself of the frontier. Rather than an external barrier (or even a permeable membrane), an emerging model of urban life relies on a perpetual "hereness" that eludes the possibility of a "there." I refer to that practice and perception as omnitopia, a term that requires further explication.

## Omnitopia and the End of the Frontier Myth

Omnitopia enacts a structural and perceptual enclave whose apparently distinct locales convey inhabitants to a singular place. Etymologically, the term reflects a neologism of the Latin *omni* (all; "in all ways or places" or "of all things") and the Greek *topos* (place).[9] Omnitopia reflects a different kind of place than its cousins, utopia and heterotopia. Utopia refers to a fictional site coined in 1516 by Thomas More, an impossible place (not-place) conceived to critique the social conditions of its author. Heterotopia refers to Michel Foucault's (1986) formulation of a paradoxical place that encloses and relieves social tensions.[10] While utopia is strictly limited to the imagination of one who critiques social life, heterotopia refers to a complex site such as a theme park, graveyard, or motel that acts as a safety value to affirm social order, even while admitting multiple and contradictory narratives. While its lineage emerges from utopia and heterotopia, omnitopia differs from these two concepts in one essential way. Rather than evoking an imaginary locale or a distinct place, omnitopia reflects the practice and perception of multiple locations being accessed through a single site. Put another way, omnitopia posits the shrinking of human geography into a singularity in which any place becomes every place and every place is the same.

As a framework, omnitopia continues to evolve past its initial limitations. The term was introduced first in an essay on airport architecture (Wood, 2003a) before its application to *The Simpsons*' depiction of Springfield (Wood & Todd, 2005), early strategies employed by "Mom and Pop" motels to craft a sense of place along the highway (Wood, 2005a), and the construction of tourist enclaves in sites such as Las Vegas (Wood, 2005b). Through a process of development and delineation enabled by these latter three essays, omnitopia has revealed itself as being far more complex than the homogeneous and interchangeable gates of the international airport. Recalling his 1984 circumnavigation through various airport enclosures, essayist Clive James noted that "the world had grown smaller without necessarily becoming bland" (1989, p. 362). Cracks and fissures inevitably cut through the monolithic façade. A manifestation of late modern capitalism, omnitopia flows like mercury, a silvery liquid that transforms natural and artificial terrains into reflections of its presence. Though often toxic, the strategies of omnitopia manage to craft a paradoxical place that is local and universal, personal and alienating. Gazing beyond the mere reflection of omnitopian practices, we must excavate fundamental practices. To that end, this essay offers a rereading of *Dark City* as an exemplar of five omnitopian strategies: dislocation, mobility, conflation, fragmentation, and mutability.[11] As discussed later, these practices reflect the evolving relationship of the frontier and urban life.

## Dislocation

Dislocation removes a site from its surrounding environment. In doing so, this strategy constructs a separate enclave that offers a synecdoche of the world marked by a limited range of experiences. Dislocation may be best illustrated by those iconic sites of late modern capitalism: the airport, the enclosed shopping mall, and the casino. For practical reasons, of course, each site must be separated from the external world. Airports require security fences; enclosed shopping malls protect their inhabitants from the elements, and casinos seek to focus their consumers' attentions on the gambling opportunities within. But in a larger sense, each of these attempts a kind of cocooning that reflects a new urban aesthetic: the power to isolate ourselves from others. By way of further illustration, consider the ubiquitous iPod (and other types of personal media devices), which allows its users to create an even more snug enclosure, an aural enclave. With this device, people carry with them their "world" of media—images, songs, movies, data files—and are able to disassociate themselves from the world "outside." Offering a world within a world, dislocation removes the locale from its human and physical geography, constructing a shopping mall-like enclosure. In *Dark City*, we encounter a nearly perfect example of dislocation, given that city residents are literally separated from their world, with no view of the frontier that divides them.[12]

*Dark City* presents an American metropolis where the sun never shines. But the city is, in fact, a self-contained enclosure where aliens known as "the strangers" experiment upon abducted people in hopes of understanding the unique capabilities of human beings.[13] This truth is hidden from the human inhabitants. But one, a character named John Murdoch, begins to recognize the unreal nature of his urban habitat. He wakes up in a gloomy hotel room and discovers a murdered prostitute, a bloodstained knife, and an unknown face in the mirror. Murdoch cannot remember his name or recall what led him to this place. He does not think he killed the prostitute, but he cannot be sure. This opening scene launches the film's plot, in which Murdoch searches the city for answers while police detectives search for him. In their stumbles through the urban scene, characters discover the detached nature of their surroundings.

Throughout the film, characters are dislocated from their place of origin and confined in a seemingly impenetrable labyrinth marked by two types of imagery: spirals and aquarium enclosures. Spiral imagery includes the film's opening credit graphics, the twisting carve caused by the murder weapon, a doctor's circular maze, and several pertinent pieces of narrative. In one, the chief inspector speaks to a detective, Frank Bumstead: "This killer's been running circles around us." Later, Bumstead spots circular mazes in the home and office of a colleague, Eddie Walenski, who has gone mad in his efforts to understand the true nature of the city: "spending time in the subway, riding in circles, thinking in circles." Throughout the film, Murdoch, Bumstead, and Walenski struggle to make sense of their world. But continually they confront signs of the spiral—spinning clocks, turning chairs, the shape and layout of the city itself.

Along with spirals, the film employs aquariums to emphasize the role of surveillance in this alien habitat. In the movie's opening scenes, Murdoch accidentally breaks a small fish bowl but rescues the fish. Later, a cop places the fish in a glass, peering at it as the strangers peer at them. At first, only Detective Walenski can see beyond the human aquarium: "They're watching us! There's no way out! God, can't you see?" When Murdoch searches for a way out of the city, he meets Walenski, who has finally abandoned any hope of freedom, choosing to leap into the path of an oncoming train. These signs of hopelessness, the spiral and the aquarium, suggest a place from which one may never escape—the impossibility of a frontier that one may pass. This place is dislocated from the larger world of choices perfectly, a floating prison with invisible walls. This is not to say we are frozen, however. Indeed, a second essential strategy of omnitopia concerns our mobility.

## Mobility

Mobility orients a place around technologies of conveyance in such a manner that the environment becomes an interconnected series of passages. We find it almost unthinkable to critique these types of places because we endlessly move through them. In this "game without frontiers," one must move in order to play. The strategy of mobility reduces, even criminalizes in some sites, the possibility to reside in one omnitopian location for an unauthorized purpose or inordinate amount of time. Rather than moments of stasis, one encounters a perpetual experience of flow. To some extent, of course, mobility constitutes an inevitable component of any urban site, particularly as the populations of people and places expand beyond the pedestrian expanse where one may walk comfortably. Even in a small town, one must practice some degree of mobility, even to visit the central park for an afternoon picnic. Omnitopian mobility, however, differs from traditional urban mobility by transforming places of stasis into sites of despair. In *Dark City*, the ability to soar over the city is limited to the strangers: literally, overseers. In contrast, the city's human population is confined to more easily monitored forms of transportation.

*Dark City* presents a dichotomous site where powerful researchers may move throughout the site at will, while their hapless subjects are restricted to vehicles whose trajectories are confined to an urban maze. A quick glimpse at Murdoch's driver's license reveals that he lives in "East City." Later, he peers at a phone book whose listings are oriented around conspicuously generic sections North, South, East,

West, The Place, The Crescent, and similar locales. These place names suggest an urban geography oriented around a bird's-eye-view map that mirrors a laboratory maze shown early in the film. The city's primary modes of mobility, trains and subways, enable inhabitants to move from section to isolated section without ever glimpsing the complex whole. Even when Murdoch searches a subway map for direction, he finds only a jumble of crisscrossing lines that depict no discernable pattern. Certainly, some inhabitants possess automobiles that, presumably, afford the freedom to transform the city into a more coherent personal narrative. However, the strangers can stop all modes of transportation whenever they choose, rendering human movement subject to their will. Moreover, the strangers are not limited to the scripted and mapped transportation modes of the city's human inhabitants: roads and train lines. Instead, they fly above the city, surveying its passages from an unobstructed view. As a result, the strangers possess a perception of the city as an omnitopian whole while the humans see only parts. This is not to say that human beings are entirely isolated in cells, though. Indeed, while the inhabitants of *Dark City* (and most inhabitants of "real" urban environments) may never witness the structural coherence of distinct locales, they often spot the playful collaboration of façades that construct a surface-level cohesion to otherwise confounding narratives. This strategy may be labeled conflation.

## Conflation

Conflation merges disparate settings into a singular whole. This strategy, therefore, arises both from the all-seeing gaze that peers over a landscape of overlapping narratives and from the pastiche design of architectural forms that disdain singular stories. Even so, this strategy should not be confused with the mere "bird's-eye-view" afforded from locations of great height such as touristic lookout towers. Instead, conflation works through the hyperspatial condensation of impossibly distinct places, peoples, and times.[14] Moreover, this strategy should not be seen as producing a genuinely coherent understanding of the structural "reality" of place. Conflation works only at the surface level in a manner similar to the experience of visiting an amusement park. There is no frontier in the conflated world because there is no "outside" referent. All becomes jumbled together. In a similar manner, *Dark City* offers soaring views over conflated urbanity whose disparate forms and referents are both impossible and increasingly familiar.

In *Dark City*, architectural forms and historical periods blur in an urban simulacrum that intentionally mirrors previous films such as *Nosferatu*, *Metropolis*, and *Blade Runner*. Recalling Baudrillard's (1994) description of simulacrum as a measure of distance from the real we might find within these films a simulacrum of simulation in which:

> [T]his projection is totally reabsorbed in the implosive era of models. These models no longer constitute either transcendence or projection, they no longer constitute the imaginary in relation to the real, they are themselves an anticipation of the real, and thus leave no room for any sort of fictional anticipation—they are immanent...The field opened is that of simulation in the cybernetic sense, that is, of the manipulation of these models at every level. (p. 122)

Characters occupy the same kinds of bottom-dwelling locales, stalking each other beneath the same types of moving trains and nightmare buildings. Beyond these cinematic references, *Dark City* confounds its viewers with conflicting architectural styles and with intentional anachronisms: a steampunk conflation of gargoyles and automats.[15] Buildings range from heavy Gothic to streamlined moderne to post-war neo-brutalism, mixed with Greek columns and Roman arches; these amalgamations render the city "both familiar and strange at the same time" (Gerlach & Hamilton, 2004, p. 120). Thus, in one scene, Murdoch converses with a man wearing a nineties-era hooded sweatshirt in a forties-era subway. Elsewhere, we survey a broad chronology of automobiles, from thirties-era roadsters to fifties-era sharkfins to seventies-era behemoths. We spot sixties-era "mop top" haircuts advertised in a Depression-era barbershop. Even the alien technology confounds our temporal sensibility, illustrated by a high-tech syringe that resembles a piece of Victorian bric-a-brac. Milner (2004) explains, "If it looks quite literally fantastic, then this is because that is what it is: a phantasm, an impossible city" (p. 269). At one point, Murdoch confronts one of the strangers who explains why the city's lack of coherent narrative serves to affirm the power of the overseers:

> Mr. Hand: There's no escape. The city is ours. We made it.
> John Murdoch: What are you talking about?
> Mr. Hand: We fashioned this city on stolen memories. Different eras, different pasts, all rolled into one.

Initially, one may be inspired to label this city as postmodern, that playfully ambiguous mélange of winking references and knowing asides. However, the locale of *Dark City* is far more ominous. Slipping between the shadows and fluorescent lights, the omnitopian city invokes multiple narratives while affirming only one, that of the architects. Even so, the resulting place and personality of its people are not integrated but instead fragmented: a confusing hodgepodge of momentary experiences with no recognizable context.

## Fragmentation

Fragmentation offers a complementary strategy to conflation. As the omnitopian environment gathers disparate surface-level images into the same perceptual experience, it simultaneously fractures itself into seemingly distinct constructions, often defined by internal rather than external order. This strategy eschews the utopian homogeneity of the airport, the planned community, and the office park.[16] As a practice of late modern capitalism, fragmentation perfects the lessons of designers and planners who have learned from Las Vegas by crafting abrupt spaces of apparent novelty and variation that remove the coherent narrative that risks critique. The frontier offers a battle line, but fragmentation removes the axis that one might attack. In this manner, fragmentation works to splinter the sweeping historical narrative, crafting incongruent shards of the present, everywhere complete and distinct. We find fragmentation in our visits to a fast food restaurant whose worker speaks to people at the counter and in their cars via a microphone and headset, rendering both interactions automated and alienating. In a similar manner, inhabitants of *Dark City* experience fragmentation of selves, relationships, and places.

In the film, the self is fragmented to such an extent that, in their natural states, all people are strangers even to themselves. By way of illustration, we might return to John Murdoch's awakening in the hotel. Unaware of his name or history, he exits and wanders down a dark alley (though all roads in this film resemble dark alleys). Staring at a store window, Murdoch finds his reflection imposed over two mannequins and "tries on" several potential selves:

> John Murdoch: What's your name? Justin? Jerry? No. My name's Jason Murdoch, John Murdoch, Jake Murdoch, How's it going? Hi. Jack Murdoch.

[He laughs in absurdity and frustration, before realizing that a street sweeper is staring at him]
John Murdoch: Hi. Gotta get a grip on myself.

When Murdoch encounters the strangers for the first time, he questions their identities even while they give him reason to question his:

John Murdoch: Who are you?
Mr. Hand: We might ask the same question, yes?

A self with no awareness of history becomes a malleable object when manipulated by the strangers. Yet the strategy of fragmentation is hardly limited to the individual.

Murdoch's visit to an automat provides an effective demonstration of relational fragmentation, while mirroring this strategy on an architectural level. The automat presents a series of discrete glass-covered containers, each containing plates of food. Patrons enter a singular environment but face an array of distinct choices. Blackmore (2004) describes the automat in *Dark City* as "a crude memory bank" for the city's ever-shifting realities (p. 25). Searching for his lost wallet, Murdoch speaks to a fleeting image of a face as the automat attendant places new food items in individual cells:

Attendant: You left your wallet here, buddy.
John Murdoch: When, when, when did I do that?
Attendant: When you was last here.
John Murdoch: When, when was that?
Attendant: When you left your wallet. You expect me to remember?

The attendant says nothing more but places the wallet inside one of the cells, forcing Murdoch to make a purchase. Like our fast food encounters with fragmented employees, Murdoch speaks to an atomized object, not an integrated person.

Beyond relational fragmentation, the city itself is split between a dark presence and a light absence. For reasons that inspire considerable debate among fans of *Dark City*, the inhabitants recall one place where the sun shines and the labyrinth opens to an open shore with no walls, where the fish are free from the aquarium: a place called Shell Beach. This idyllic locale resides in fragments: dreams, distant memories, and tantalizing souvenirs. Indeed, the film offers multiple moments in which the dark city is fragmented with the introduction of these souvenirs: postcards, photographs, scrapbooks, and billboards depicting the sun over a beach. Yet none of these referents reflects the physical reality of the city. While visiting his Uncle Carl, Murdoch discovers mementos from a "childhood": shells, marbles, photographs, and then a scrapbook: *Guide to Shell Beach*. He opens the book expectantly, in search of an answer to the mystery, but all the pages are blank. He searches the city for a passageway to Shell Beach, but finds only images, posters over a brick wall.[17] The pieces to the puzzle never fit. One primary reason for this phenomenon is that they perpetually change.

## Mutability

Mutability allows for the transformation of a place for multiple purposes, which range from playful manipulation to coercive control. Mutability relates to mobility through its negation of fixed environments. However, we do not simply move through omnitopian locales that remain relatively constant. Instead, we encounter places, objects, and even people that move, change, and alter themselves in frequently unforeseen manners. All that is frontier melts into air. Expanding beyond this somewhat literal notion of mutability, we find environments throughout urban life that adapt themselves to changing needs. Thus, we recognize that an airport gate leads to one location in the morning, another in the afternoon, and yet another in the evening. Of course, *Dark City* presents an even more potent example of architectural transformation with its depictions of places and people that change according to the whims of the strangers.

At first, *Dark City* presents its viewers only tantalizing glimpses of the true nature of this place. Some of these images are strictly symbolic—an automat attendant replacing a banana with a bowl of green jello—while others are more literal: places that exist only as mutable constructs. Once more, let us consider Shell Beach, that beacon of half-remembered childhood and sunny days, a mystical place that recedes into a cloudy distance whenever anyone tries to determine its actual location. Throughout the film, Murdoch searches for Shell Beach. At one point, he enters a taxi and spies a souvenir, a snow globe from the beach, and he asks for directions:

> John Murdoch: Hey, you happen to know the way to Shell Beach?
> Taxi Driver: You're kidding! Me and the missus spent our honeymoon there. All you gotta do is take Main Street West to—or is it the Cross. You know, that's funny, I, I can't seem to remember if it's Main Street West or the Crosstown.

The driver cannot remember the location of Shell Beach because its location is not fixed. Rather, it constantly changes. Beyond fluid directions, *Dark City* confounds its characters (and viewers) with shifting locales and disconnected motives, such that detective Bumstead eventually despairs, "I have this jigsaw puzzle in front of my face, and every time I try to rearrange the pieces it still doesn't make any sense!" Bumstead demonstrates more insight than he can imagine, at first. The city is a giant puzzle whose pieces are not only mismatched but also altered perpetually.

In one of the most unique contributions made by *Dark City* to urban imagery, we learn that the strangers have not built a fixed habitat for their captives but rather have crafted a mutable place that grows and changes every night. Their power to alter the city is called "tuning," a nightly transformation in which buildings twist and contort themselves, adapting themselves to new narratives implanted by the strangers. Higley (2001) offers an apt description, noting how "the revisions that the aliens make to the urban cityscape come torquing up out of the ground like great architectural corkscrews; buildings collapse as others emerge and battle with one another at the end" (p. 10). During the "tuning," all residents are put to sleep, only to awaken in new settings with new memories. But John Murdoch has awakened in the middle of the tuning and slowly learns that he too has the power to alter the city and adapt to the changes wrought by the strangers. Thus, in a central scene of mutability when Murdoch flees his captors, he jumps onto a growing chimney that becomes, in effect, an elevator. He eventually learns to tune the city himself, and in doing so presents both a grave threat and long-awaited opportunity for the strangers.

As with physical architecture, the personalities who are the city's inhabitants endure changes every midnight when the clocks stop and the strangers go to work. As a result, the inhabitants of *Dark City* contain disparate histories and psychological profiles as the strangers experiment with their captives, literally crafting their personalities with the aid of a human collaborator, Dr. Daniel Schreber, used by the strangers because of his insights into human psychology. Thus, John Murdoch has been imprinted with the personality of a mass murderer, drawn from dozens, perhaps hundreds of human memories, in order to provide a test case for the strangers. Schreber explains, "This city, everyone in it, is their experiment. They mix and match our memories as they see fit, trying to divine what makes us unique. One day, a man might be an inspector. The next, someone entirely different." Indeed, part of the film's pleasure is the chance to spot characters that have undergone revision, such as when a night desk clerk becomes transformed into a newspaper vendor. Like the city, people have become containers for strange combinations, the "killer" who rescues a dying fish, the "hard-boiled detective" who pines for his dead mother. Schreber later says to Murdoch, "Your entire history is an illusion, a fabrication, as it is with all of us." Aided by Dr. Schreber's chemical amalgamations, citizens of *Dark City* wake up to new personalities. This mutation of places and people reflects an omnitopian sensibility that contains a peculiar pleasure, a freedom from consequence. Yet, at the same time, we experience alienation from environments and other human beings. A hero like John Murdoch learns eventually to edit the city for his own purposes, but the rest of us appear merely to pass through its mutating regime, disoriented but entertained.

## From Old Frontier to No Frontier to Artificial Frontier

These strategies—dislocation, mobility, conflation, fragmentation, and mutability—reveal practices of omnitopia. Dislocation detaches a place and its people from the larger world, constructing an enclosure whose membranes may be relatively permeable but are nonetheless difficult to transgress without authorization. Mobility enables, even requires, movement through distinct locales within the enclosure, representing a strategy that diminishes the potential for reflection and critique. Conflation works to collapse varied environments and narratives, even temporal ones, into a singular arcade: crafting an illusion of completeness. Fragmentation works simultaneously to craft abrupt moments of disjuncture, distinct sites of "now," such that each grand view and conflated vision becomes isolated and alienating. Mutation draws from each of these strategies to enact states and perceptions of constant change, removing the physical or temporal coherence necessary to mount a meaningful critique. Thus, while we imagine ourselves in the position of John Murdoch, the knowing hero who transforms the city into his personal playground by film's end, we risk adopting the altered guise of Emma Murdoch who resurfaces after the final "tuning" as a human stranger with a random identity and no clue how she got there. In this "small world," and in others throughout the "real" world, Emma's fate seems more likely than John's.

What remains is for us to investigate whether such a fate is, indeed, inevitable. In this way, we exit the theater and enter the modern world once more. Modern urbanity envisions a world of grand narratives, order, and precise placement. As evidence, one might recall nineteenth-century Haussmannization of Paris[18] or twentieth-century "urban renewal" in the United States, perfectly illustrated by the transformation of New York's Times Square. Both illustrate the perfect and totalizing gaze of the strangers who present themselves to us as bureaucrats. In the midst of such vast enclosures and grand boulevards, individuals bump up against each other like billiard balls, alienated from one another. Objects of the modern world become abstract. As both a consequence of this condition and a reaction to it, omnitopia enacts the totalizing gaze and ubiquitous place of modernity, creating an enclave of order amidst a desert of disorder. Of course, we are not entirely enclosed in this arcade. In its conclusion, *Dark City* offers a tantalizing glimpse of how technology may be used to reconstruct the frontier through the heroic efforts of the individual who frees himself from the constraints of the city. Initially, this "new frontier" offers an affirming vision, an optimistic pinnacle of human will to power. However, as I explain, *Dark City*'s depiction of the new frontier rests on a troubling artificiality.

While I would prefer to withhold the film's conclusion from readers in hopes that some who have not yet seen *Dark City* might enjoy learning its secrets for themselves, I am compelled to explore the film's surprising resolution.[19] Although cast as opponents by the strangers' designs, John Murdoch and Inspector Bumstead eventually join Dr. Schreber, forcing him to reveal the purpose of their plans. For Murdoch, this quest ends at Shell Beach, that mythical frontier between city and nature that no one can seem to reach. Schreber warns that Murdoch can indeed visit Shell Beach but that he will not like what he finds. Refusing to take heed, Murdoch and Bumstead journey to the edge of the city in search of the strangers' secret. Surely at the water's edge they might escape their captors' grasp. But once they arrive, they find that Shell Beach is merely an advertisement stuck on a brick wall. There is no "outside" to the city. Nevertheless, Murdoch and Bumstead grasp hammers and begin to strike at the wall, unaware of what awaits them. With blow after blow they pummel at the façade, ripping off the advertisement and forging a crack in the wall before discovering the film's secret at last. The city is, in fact, a vast spaceship turned always away from the sun. There is no end to the enclosure but that ultimate desert of space—a new frontier.

As the film reaches its climax, Schreber teaches Murdoch to harness his mental powers to control the machines built by the strangers, to control the "world" in which they are enclosed. This transformation is enhanced by the injection of a syringe loaded with artificial memories so that Murdoch believes that he has been preparing to control the city from birth. An elegant solution, yes, but a troubling one. As Murdoch uses his previously dormant psionic abilities to battle with the strangers, he literally ascends from the city. At this point the film resembles a comic book in its use of iconic imagery and depiction of the otherworldly struggle between Murdoch and the lead stranger. But soon enough Murdoch prevails and returns to the surface of a city that he and the stranger wrecked during their combat. The city is his to rebuild or to alter, as his imagination deems proper. In a manner suggestive of the power possessed by David Bowman as the Star Child in Arthur C. Clarke's novelization for *2001: A Space Odyssey,* Murdoch uses his mental powers to turn the spaceship-city toward the sun for the first time. Then he unleashes torrents of water from the city's edge to form an ocean bordered, at last, by a sandy frontier. Murdoch cannot find Shell Beach so he chooses to build one.[20]

At film's end, the city resembles the Earth that Murdoch recognizes, complete with sun, clouds, and water. Shell Beach promises that mythical frontier between city and nature that defines both in relation to each other. But, of course, the frontier is an illusion, as is Murdoch. Both are based on false memories, cobbled together in Schreber's lab. The city resembles Earth, but it continues to float as a fake in the vastness of space. Murdoch accepts his identity, but he knows it is an amalgam of choices he did not make. The freedom of choice offered by *Dark City* is an illusion, one that can only be altered by a new stranger. In this manner, the film presents "a definition of identity as self-invention" (Marsen, 2004, p. 145). We find a model in Daniel Schreber's advice to John Murdoch: "This is the machine the strangers use to amplify their thoughts, the machine that changes their world. You must take control of it. You must make the machine yours." Leaving the film, we find ourselves faced with a similar responsibility, to take the world back upon our shoulders. Such a challenge can be overwhelming, but it can be liberating too. We have no map and, thus, are forced to make our own way.

## Conclusion

This chapter has offered a reading of *Dark City* as a means to trace the emergence, death, and renewal of the frontier in urban life. By exploring the historic (and historiographic) construction of the frontier as a counter-city, outlining how the omnitopian city rids itself of the frontier, and unpacking some implications on how individual "power" within the omnitopian confine contains the seeds of its own opposition, I have attempted to offer more than a summary of a largely unappreciated film. Rather, this chapter has promoted a vision of the mythical frontier that, like "real" boundaries, has managed to shift and mutate with changing exigencies. Like the mercury of modernity, the mythical frontier reflects us to ourselves, showing us what may be otherwise obscured by the shocks and pleasures of urban life. In the omnitopian frontier, we find the illusory space of agency that reminds us of a videogame whose "edges" are soft and sometimes impossible to discern.[21] We bump against pixelated borders in the softest of rebuke, for our choices that lie behind us seem limitless. So we turn back to the city (of bits, of dreams, of the dead) and play our roles with renewed vigor, forgetting that the world beyond our "small world" is larger than we can yet imagine.

## Notes

1 *Dark City* shares a name (and some of the tone) with a 1950 film directed by William Dieterle and starring Charlton Heston.

2 While the film drew from decidedly American imagery, *Dark City* was helmed by an (Egyptian-born) Australian director and shot in Sydney.

3 Beyond his statement that *Dark City* was the best film of 1998, Ebert's enthusiasm for the film is best illustrated by the commentary track he contributed to the DVD, a contribution that serves as a master class in film analysis and criticism.

4 Readers now can be forgiven for being underwhelmed by this thesis; it seems so ordinary and so conventional. However, this response merely affirms the power of a new idea to seep into the academic milieu before infecting social, economic, and political sensibilities until it becomes the lens through which common sense may be judged. Turner's frontier thesis offered no less than a revolution in American thought, invoking the westward-bound frontiersman rather than the European ancestor as the defining mythical figure of the new nation.

5 This is not to say that Turner (1920) announced the end of the nation's expansionist narrative. Indeed, despite the closing of the frontier, he assured his

readers that "the American energy will *continually* demand a wider field for its exercise" (p. 37, emphasis added). America's grand and occasionally galling adventures in the century that followed would bear that thesis out. Indeed, Ronald H. Carpenter (1990) traces some of the implications of that expansionist ethic.

6 Dale (1941) wrote, "there is still a carry over of the frontier speech to a modern social order that never knew the life out of which this speech grew" (p. 362).

7 In his essay, Clark (1942) weaves a narrative that is typical of scholarship about the frontier, the unfolding American drama as seen through the westward march of trailblazing pioneers. Rather than attempt a similar survey, I repeat his: "the true *West*, the real *frontier* in American history, was a continually advancing line. At first it lay along the edge of the Atlantic sea coast, then it moved inward to the Piedmont, to the Alleghenies, to the Ohio country, and thence across the river to the plains and again to the mountains. But always it was the *West*, the *frontier*, and always it was distinguished by its disregard for established institutions and by its new and genuine democracy, both of which grew out of the exigencies of the environment" (p. 282).

8 As Carpenter (1977) explains, "A statement intended to alter the course of American historiography became instead the rhetorical source of a mythic, national self-conception" (p. 128).

9 The etymology of *omni* is quoted from the *Oxford English Dictionary*.

10 See also Soja (1995, 1996, 1997) and Wood (2003b, 2004).

11 This analysis reflects a particular extension of a recent typology of omnitopian strategies (Wood & Todd, 2005). Most notably, the present work provides a more precise relationship between fragmentation and conflation. Also, this analysis removes commodification from the list of strategies. Certainly, one senses the role of commodification in *Dark City*, most notably when the hotel desk clerk announces, "Only thing that makes you a guest in this joint, pal, is cash on the barrelhead." However, on reflection, I propose that commodification (along with atomized interactions described in Wood, 2003b) resides as an effect of omnitopia—related to fragmentation—rather than a strategy of its design.

12 A reader of an earlier draft noted the presence of the river in the first scene of this chapter, potentially representing a frontier between the city and the non-city beyond. However, the scene is framed in such a way as to show only the city with no contrast, no outside. The river is a setting to this scene but not a divider to the city. Here I recall a scene from the movie *Pleasantville* that illustrates this paradox. In a classroom, a visitor to the city asks, "What's outside of Pleasantville?...What's at the end of Main Street?" Her question results in chuckles and confusion. Inhabitants in this city recognize the existence of a road, just as inhabitants of *Dark City* recognize the river. But in *Pleasantville*, "The end of Main Street is just the beginning again." Similarly, in *Dark City*, the river cannot lead to the "outside," until the film's climax.

13 Not too surprisingly, Gerlach and Hamilton (2004) note that *Dark City*'s depiction of urban "strangers" applies "as easily to Georg Simmel's work on the city as it does to Alex Proyas' 1998 film" (p. 115).

14 Along with our *Dark City* exemplar, we might consider other media examples of conflation, such as the fifties-themed restaurant, Jack Rabbit Slims, in Quentin Tarantino's *Pulp Fiction* (a film whose car-tables were inspired by the 1968 Elvis Presley film *Speedway* and its Hangout Club). We can also recall George Miller's 1998 *Babe: Pig in the City,* whose convergence of architectural icons such as the Empire State Building, the Sydney Opera House, and the Golden Gate Bridge created a charmingly paradoxical metropolis. Along with these films, "tiny towns" provide another parallel to the synecdochic enclosures of urbanity found in omnitopia. Examples around the United States include Tiny Town U.S.A. in Hot Springs, Arkansas; Roadside America in Shartlesville, Pennsylvania; the Ave Maria Grotto in Cullman, Alabama; the Crossroads of America in Bethlehem, New Hampshire; and the DeLorme "Eartha" globe in Yarmouth, Maine. A delightful international example may be found at Cockington Green in Canberra, Australia.

15 Francis Mason offers a useful parallel with his essay on British "pop" sci-fi. Mason (1996) describes "the nostalgic citation of the past...a selective process where the 'better' (modern) things from twentieth-century history can be referenced and then transposed on to this future space, particularly style and fashion" (pp. 30–31).

16 A quintessential example of office park homogeneity may be found in the 1999 Mike Judge film *Office Space*. Even so, this film also manages to portray a kind of heterotopian locale in its depressing depiction of Chotchkie's, a satire of restaurants such as TGI Friday's (and its *former* slogan). While heterotopian in its "playful" overlap of referents, the theme restaurant becomes omnitopian when one experiences the sense of being "in the same place," whether entering "this" restaurant or "another" one. The rhetoric and perception of *ubiquity*, even more than its fake pastiche of elements, marks the omnitopian enclosure. Consider the Friday's slogan: "In here it's always Friday."

17 As a reviewer of an earlier draft of this manuscript suggests, Shell Beach demonstrates a kind of fragmentation on a deeper, almost existential level: a divide between where you "are" and where you can "not be."

18 Robert Fishman (1987) explains how the mid-nineteenth-century reconstruction of Paris sought to erect a modern capital city. With the approval of Louis Napoleon, Baron Georges-Eugène Haussmann transformed the French capital by cutting "wide, straight boulevards through the maze of narrow streets, buying up and demolishing any buildings that stood in the way. These boulevards were to create the crucial means of communication that would finally unite Paris into a great city" (p. 112).

19 Many aficionados of the film warn first-time viewers to turn down the audio until they see Dr. Schreber looking at his watch. This low-tech solution rids the film of its heavy-handed exposition that was inserted, one may assume, to assuage the fears of studio producers who fretted that audiences would not tolerate the film's deliberate unraveling of mystery.

20 Murdoch-as-urban-planner resembles the "real" ones we see around the world who draw from media-fed fantasies to transform the undulations of natural topography into a container of our desires. Thus, middle-class backyards gain waterfalls and desert cities bloom with golf courses.

21 The nature of this illusion is illustrated by a commercial that, as much as *Dark City* (and perhaps inspired by it), illustrates the experience of omnitopia: the award-winning 2003 Volkswagen "Bubble Boy" commercial. In this spot, a corporate drone passes through a series of bland enclosures before spotting a car outside a plate glass window, a means of freedom. Walker (2003) writes, "The kid is likable, so we're sort of rooting for him, but he seems trapped, somebody stuck in a huge transparent cage, looking for signs of life out there to latch onto" (¶3). For the "bubble boy," the car represents freedom. But, given the totalizing enclosure shown in this commercial, it is hard to imagine where he could possibly go.

# References


Baudrillard, J. (1994). *Simulacra and simulation.* Ann Arbor: University of Michigan Press.

Blackmore, T. (2004). High on technology—low on memory: Cultural crisis in *Dark City* and *The Matrix. Canadian Review of American Studies, 34* (1), 13–54.

Carpenter, R. (1977). Frederick Jackson Turner and the rhetorical impact of the frontier thesis. *Quarterly Journal of Speech, 63* (2), 117–129.

Carpenter, R. (1990). America's tragic metaphor: Our twentieth-century combatants as frontiersmen. *Quarterly Journal of Speech, 76* (1), 1–22.

Clark, R. (1942). The influence of the frontier on American political oratory. *Quarterly Journal of Speech, 28* (3), 282–289.

Dale, E. (1941). The speech of the frontier. *Quarterly Journal of Speech, 27* (3), 353–363.

Fishman, R. (1987). *Bourgeois utopias: The rise and fall of suburbia.* New York: Basic Books.

Foucault, M. (1986). Of other spaces. *Diacritics, 16,* 22–27.

Gerlach, N. & Hamilton, S. (2004). Preserving self in the city of the imagination: Georg Simmel and *Dark City. Canadian Review of American Studies, 34* (2), 116–134.

Hayles, N. & Gessler, N. (2004). The slipstream of mixed reality: Unstable ontologies and semiotic markers in *The Thirteenth Floor, Dark City,* and *Mulholland Drive. Publications of the Modern Language Association of America, 119* (3), 482–499.

Higley, S. (2001). A taste for shrinking: Movie miniatures and the unreal city. *Camera Obscura, 47* (2), 1–35.

James, C. (1989). Around the world in one pair of shoes. In C. James (Ed.), *Snakecharmers in Paris: Essays 1980–87* (pp. 357–366). London: Picador.

Marsen, S. (2004). Against heritage: Invented identities in science fiction film. *Semiotica, 152* (1.4), 141–157.

Mason, F. (1996). Nostalgia for the future: The end of history and postmodern "Pop" T.V. *Journal of Popular Culture, 29* (4), 27–40.

Milner, A. (2004). Darker cities: Urban dystopia and science fiction cinema. *International Journal of Cultural Studies, 7* (3), 259–279.

Rushing, J. (1983). The rhetoric of the American western myth. *Communication Monographs, 50* (1), 14–32.

Rushing, J. (1986). Mythic evolution of "The new frontier" in mass mediated rhetoric. *Critical Studies in Mass Communication, 3* (3), 265–296.

Rushing, J. (1989). Evolution of "The new frontier" in *Alien* and *Aliens*: Patriarchal co-optation of the feminine archetype. *Quarterly Journal of Speech, 75* (1), 1–24.

Soja, E. (1995). Heterotopologies: A remembrance of other spaces in the Citadel—LA. In S. Watson & K. Gibson (Eds.), *Postmodern cities and spaces* (pp. 13–34). Oxford: Blackwell.

Soja, E. (1996). *Thirdspace: Journeys to Los Angeles and other real-and-imagined places*. Cambridge, MA: Blackwell.

Soja, E. (1997). *Postmodern geographies: The reassertion of space in critical social theory*. New York: Verso.

Tripp, D. (2005). "Wake Up!" Narratives of masculine epiphany in millennial cinema. *Quarterly Review of Film and Video, 22* (2), 181–188.

Turner, F. (1920). *The frontier in American history*. New York: Henry Holt and Company.

Walker, R. (2003). Freed by the beetle: Volkswagen rescues a cubicle slave. *Slate*. Retrieved November 2, 2006, from http://www.slate.com/id/2076322/

Wood, A. (2003a). A rhetoric of ubiquity: Terminal space as omnitopia. *Communication Theory, 13* (3), 324–344.

Wood, A. (2003b). The Middletons, Futurama, and Progressland: Disciplinary technology and temporal heterotopia in two twentieth century New York World's Fairs. *New Jersey Journal of Communication, 11* (1), 63–75.

Wood, A. (2004). Managing the lady managers: The shaping of heterotopian spaces in the 1893 Chicago Exposition's Woman's Building. *Southern Communication Journal, 69* (4), 289–302.

Wood, A. (2005a). "The best surprise is no Surprise": Architecture, imagery, and omnitopia among American mom and pop motels. *Space and Culture, 8* (4), 399–415.

Wood, A. (2005b). "What Happens (in Vegas)": Performing the post-tourist flâneur in New York and Paris. *Text and Performance Quarterly, 25* (4), 315–333.

Wood, A. & Todd, A. (2005). "Are we there yet?": Searching for Springfield and *The Simpsons*' rhetoric of omnitopia. *Critical Studies in Media Communication, 22* (3), 207–222.

# Critiquing the Myth and Looking to the Future: Slayers, Starfleet, and the Schwartz

•CHAPTER EIGHT•

# *Buffy the Vampire Slayer* and the Quest Story: Revising the Hero, Reshaping the Myth

*Tanya R. Cochran and Jason A. Edwards**

You think you know. What's to come, what you are. You haven't even begun.

—Tara, speaking for Sineya, the first Slayer (Whedon, 2000b, "Restless")

When Joss Whedon writes for film or television, he actively seeks, by his own admission, to subvert established genres. To upend teen horror, for example, Whedon makes the stereotypical small, blonde, beautiful and first-to-die-violently young woman of the horror film the hero of his movie and subsequent TV series *Buffy the Vampire Slayer* (*Buffy*); he intentionally attempts to overturn not only established genre but also social order. As an extension of his cultural work, we argue that Whedon has done his part through *Buffy* to revise and reshape what Joseph Campbell considers *the* human story, the myth of the hero's quest.

The hero's journey or quest story can be found in the mythologies of all peoples; it crosses times, continents, and cultures. Most easily and visibly located in popular fiction and film, the phases of the quest surface over and over again, giving us such recent examples as the multiple incarnations of *Star Wars, The Matrix*, *Harry Potter*, and *The Lord of the Rings*. Though one might argue it has fewer clear

* Tanya R. Cochran (M.A., University of Tennessee at Chattanooga) is Assistant Professor of English at Union College. Dr. Jason A. Edwards (M.A., Minnesota State University-Mankato; Ph.D., Georgia State University) is Assistant Professor of Communication Studies at Bridgewater State College.

cases (obviously, *Star Trek*), television also draws on the hero's passage. Every season of *Buffy* (1997–2003), for instance, has an embedded journey, a story arc. Particular villains, or "Big Bads," dominate each season, so episodes do not stand alone. To experience the true essence of *Buffy*, viewers must watch in sequence, must journey with the hero and her companions. Also, taken as a whole, the seven completed seasons focus on Buffy's personal development as a woman and Slayer, a sort of "coming into one's own." In other words, both as individual seasons and as a whole series, *Buffy* gives viewers a hero who both lives out a personal quest and works to rid the world of evil.

Many scholars have noted how *Buffy* draws upon the hero's journey (Bowman, 2002; Buttsworth, 2002; Early, 2002; Wilcox, 2005); at the very same time it performs the quest, however, *Buffy*, like any Whedon creation, aims to misbehave. For example, Bowman (2002) argued that *Buffy*'s most notable tie to myth is the prominence of the hero within the story. According to Bowman, Buffy herself subverts the typical image of the quest story hero (one who is male, a loner, and excessively aggressive) because she is not merely a woman but a woman who has stereotypically distinctive feminine markers (i.e., she is thin, "pretty," at times self-absorbed, and at others infatuated with boys) (Labre & Duke, 2004; Owen, 1999);[1] in fact, Buffy often queries aloud why her life cannot be or is not more "normal." These "normal" characteristics for a teenage girl make Buffy an antithetical quest hero.[2] Both Buffy and *Buffy*, then, revise the heroic image. We argue further that *Buffy* has significantly undermined that image in an additional way. Typically, the journey of the hero begins with a welcoming of companions; a collective mindset is established. However, by the end of the journey, the hero completes the task alone; the cohorts have either abandoned the quest or been forced to stay behind. By giving her friends talent and depth of character comparable to Buffy's and refusing to let them fall by the wayside on the quest, the series rewrites its way into modern mythology, a revision and "misbehavior" that makes *Buffy* journey-shattering. As a result of Whedon's deliberate emphasis on the characters' development, *Buffy* offers what we call the equitability of the hero. In other words, heroic responsibility, position, and power are shared communally among Buffy, her friends (specifically Willow Rosenberg and Xander Harris), her family (Dawn), and all potential Slayers.[3] To explicate our argument, we proceed as follows: first, we provide a theoretical overview of myth, the quest story, and the hero; we perform a critical reading of the series; we discuss our reading in light of other scholars' work; and finally, we draw implications from our analysis.

## Theoretical Overview: Myth, the Quest, and the Hero(es)

Oh, I'm beginning to understand this now. It's all about the journey, isn't it?

—Giles (Whedon, 2000b, "Restless")

Myth is a common form of discourse. In fact, Campbell (1949) went so far as to argue that some myths and their functions cut across all cultures and societies. At their most basic level, myths are narratives that contain a plotline with characters that constitute heroes, villains, and bystanders. Yet not all narratives are myths. A narrative of the happenings of one's day, for example, is not mythic in nature because myths involve stories that are engrained in the specific political and social culture of a particular society, stories that articulate the society's beliefs, dilemmas, and values (Rushing & Frentz, 2005). Myths, then, offer us a way to frame the reality of our situation. When we humans use them, our myths function in a variety of ways, two of which are most important to this discussion.

Myths help us comprehend the world around us (Nimmo & Combs, 1980). At times, both individuals and communities are struck by some form of disorder: a natural disaster, an attack by another nation, a specific illness, a downturn in the economy, or any other disturbance to the regularity of life. It is then myths develop or are invoked to offer a means of coping. The world is often too complex to grasp—too much information, too many people, too many countries, simply too many factors to manage all at once. Myths give our world structure so that individually and collectively we can better realize the opportunities available to us, the challenges that may pose a threat to our success, and the limitations within which we must work to accomplish an objective.

A second function of myth is that it offers a sense of identification (Dorsey, 1995; Fulmer, 1986; Kluver, 1997; Nimmo & Combs, 1980). Myths are what Nimmo and Combs (1980) called "social glue" (p. 13); they work to hold a group together by providing the basis for peoples of diverse backgrounds to find common ground with each other. This common ground helps to define who "we" were, are, and will be (Edwards, 2000; Kluver, 1997; Starr, 1973). Often a myth is used in this fashion to unite a small or large group around a common ideal, one that can be expanded even further if the group accepts a specific casting of it. The myth of divine election, for instance, holds that a community collectively believes it has an exclusive place within the overall order of communities. This community is special or "chosen" and, therefore, destined for a unique mission that will demonstrate its exceptional nature over other communities (Cauthen, 2004). A sense of chosenness provides both individuals and communities a sense of identity and place.

One of the most powerful and ancient myths is the hero's quest or journey, what Campbell (1949) called the monomyth (see also Jewett & Lawrence, 1977; Norman, 1969) and what Nimmo and Combs (1980) called a "master myth."[4] This myth is a wide-ranging, overarching narrative that permeates diverse cultural artifacts, including creative literature, film, and television series. From these artifacts, it is easy to identify a method to the madness of the journey, three general phases: departure, initiation, and return. Also emerging from the myth is a central character, often a lone hero who must accomplish a particular task (Campbell, 1949).

In the departure phase, the hero learns that a mission must be accomplished in order to rescue the local community. This hero is usually from obscure and mysterious beginnings and is thrust into the role of hero because of fate or destiny. Often, though, the hero initially rejects destiny, having first to struggle with this new identity as a savior figure and the responsibility inherent in achieving the quest. It is with reluctance that many heroes embark on a journey that promises to provide the salvation, either for themselves or for their communities, that they seek. At this point, the hero moves into the second phase of the quest, one in which she or he attempts to fulfill the calling or return with a particular object. While attempting to realize the quest, the hero is often accompanied by friends, servants, and/or guides who serve as support. Support is essential, as the hero is constantly challenged by mental and physical dangers—threatening but purifying—that must be overcome. By being tried in the fire, both internally and externally, the hero proves worthy to continue with the quest and accomplish the mission. Finally, when the quest is complete, the hero returns home with the "boon." The return journey can be just as—or more harrowing than—the initial departure, for the hero is charged with getting back as quickly as possible to supply the community with the deliverance it requires for continued existence. Upon homecoming, the hero is free to serve society once more.

In this narrative, the hero is traditionally a man who has been alienated from the world with which he is familiar yet who embarks on a journey to preserve himself or, more likely, his people. As the hero begins the journey, he is often assisted by companions who provide some guidance or aid. For example, Obi-Wan Kenobi serves as Luke's guide and mentor in *Star Wars*. We see a similar pairing of characters in *The Matrix* (Morpheous and Neo) and *Harry Potter* (Dumbledore and Harry), among other films and novels. Often, though, these companions cannot continue with the hero throughout the journey; rather, the hero must face the unknown alone. Moreover, companions are usually relegated to the background of the story where they receive little to no prominence (Nimmo & Combs, 1980); they are sidekicks, at hand to satisfy the hero's needs along the way. The singular hero is so prevalent in American film and television that Jewett and Lawrence (2002) argued a gradual rise in an American superhero has occurred. This male superhero is attracted to violence and acts out the stages of the journey with tragic consequences. Characters performed by John Wayne, Clint Eastwood, and Steven Seagal offer perfect examples.

*Buffy the Vampire Slayer* shatters these classic and American heroic attributes. Rather than sidekicks who fulfill the Slayer's every need, each of Buffy's friends becomes a central, crucial figure in the series; rather than a quest that ends in tragedy, *Buffy* ends with the globalization of power (Wilcox, 2005). In her last and most intuitive effort to save the world in "Chosen," the final episode of the series (Whedon, 2003), Buffy realizes that only by empowering, all at once, every potential Slayer from Sunnydale to Shanghai (an act that Willow accomplishes through her honed magical skill as a Wiccan) can the world avoid impending destruction as well as sustain the delicate balance between good and evil. Power and responsibility, then, are evenly distributed among both friends and would-be Slayers, thereby making the supernatural more natural, the extraordinary more ordinary: the Chosen One becomes the chosen ones, each playing an essential role in the quest's realization. Emphasis on the collective suggests a refreshing way to understand how mythic heroes can be constructed and portrayed in popular culture and society at large, a way that slowly but steadily unfolds throughout each season of *Buffy*.

Early in the series, for example, Willow and Xander define themselves as Buffy's subordinates, as is clear from their self-selected title "Slayerettes." As the seasons progress and the characters mature, however, Willow and Xander, in particular, discover and exercise their own skills, gifts, and powers; they are given individual plots contained within the overall story arc; and, ultimately, they are vital to accomplishing heroic missions. Interestingly, the Slayerettes evolve into the "Scoobies," a reference to the communal, collaborative group of animated crime fighters and mystery busters. Thus, *Buffy* redefines the quest story by moving the hero's cohorts from background to foreground; the series revises the monomyth through the equitability of heroic deeds. Certainly, Buffy is the lead character; after all, the show is named for her. Especially at the outset, she is the one who grudgingly answers the call to adventure. As Playden (2001) noted, over the course of seven seasons, however, Buffy's close friends assume heroic responsibility and position just as much as Buffy does. Not only do they take on the role of heroes—including the obligatory obstacles and (mostly) certain victories—but they also reap the rewards of successfully enduring the journey. In many instances, the Scoobies take on assignments usually reserved for the Slayer alone; that they are even privy to Buffy's identity as the Slayer sets them apart. On *Buffy*, then, heroic acts are not restricted to one character; rather, they are shared among all characters, particularly the ones who become her chosen family (for an extended discussion of chosen families in *Buffy*, see Battis, 2005).[5] By emphasizing the collective nature of fulfilling *Buffy*'s interpretation of the quest story, the series establishes and celebrates equality among its characters. This sharing of the hero's characteristics, trials, and rewards goes against the orthodox construction and understanding of what it means to be heroic, a construction that limits the potential for and the potential of many new creative mythologies. To remain in such a framework is to accept and even perpetuate privileged ways of thinking about life's journey. Whedon and his team of writers and directors do not allow Buffy a solitary trek; she requires the assistance of her (extra)ordinary allies, ones who understand that "it's all about the journey" (Whedon, 2000b, "Restless").

## Critical Reading: To Shape the Myth Again

A Slayer with family and friends. That sure as hell wasn't in the brochure.

—Spike (Greenwalt & Kretschmer, 1997, "School Hard")

In its war with the Slayer and the powers of good, the First, the primeval evil, has laid waste to Sunnydale, California. So when Buffy's friends and even her sister temporarily reject her as their leader, Buffy wanders the streets until she finds a place to rest in a stranger's abandoned home. When Spike—once Buffy's enemy, then her lover, now her friend—finally finds her, he tries to convince her to take back her rightful place in the fight, to assume her role and identity as the Slayer. But Buffy refuses, reasoning, "I'm just so tired..." Her unfinished sentence suggests a number of completions: tired of being in charge, tired of being responsible, tired of being the Chosen One. In a moment of self-reflexivity more deliberate than any other in the series, Buffy realizes and articulates, "You know what? I'm still making excuses. I've always cut myself off, I've always...being the Slayer made me different but it's my fault I stayed that way. People are always trying to connect to me but I just...slip away" (Kirshner, Noxon, & Contner, 2003,"Touched"). Spike soundly disagrees:

> Hey, look at me. I'm not asking you for anything. When I say I love you, it's not because I want you, or because I can't have you. It has nothing to do with me. I love what you are, what you do, how you try. I've seen your kindness and your strength. I've seen the best and the worst of you, and I understand with perfect clarity exactly what you are. You are a hell of a woman.... You're the one, Buffy.

Spike validates Buffy, not for what she is (the Slayer) but for who she is ("a hell of a woman," "the one"). Still, fear and confusion die hard. Tears trace Buffy's cheeks as she stares down from the edge of the bed where she sits, Spike kneeling before her. "I don't want to be the one," she manages. Her rejection of being "the one" is especially significant in light of Campbell's description of the modern hero: "And so every one of us shares the supreme ordeal—carries the cross of the re-deemer—not in the bright moments of [her] tribe's great victories, but in the silences of [her] personal despair" (1949, p. 391). And yet throughout her journey, Buffy navigates the ordeal she finds herself a part of—at times the lone hero, at others the collaborative hero. Sometimes she shares the power and position of her calling, and sometimes she claims it for herself alone. This negotiation is evident from the very beginning of the series and culminates in a revision of the quest story.[6]

## The Myth of the Slayer

The first season of *Buffy the Vampire Slayer* establishes the Slayer myth, introduces characters who become Buffy's chosen family, and, via the death of the Master, lays the foundation for Buffy's revision of the myth.

Without a theme song or credits, early episodes of *Buffy* open with a voiceover that explains, "As long as there have been vampires, there has been the Slayer. Into every generation, a Slayer is born. One girl in all the world, a Chosen One. One born with the strength and skill to hunt the vampires, to stop the spread of evil." Already this declaration contrasts with what scholars have established is the purpose of myth—to unite groups of people. In the case of the Slayer myth, very few people even know about it: namely, the Watchers' Council (a group of mostly stuffy British men who "watch" Slayers by training them), the Slayer herself, and, in some cases, several potential Slayers. This declaration, this imposed identity, is what Buffy negotiates, often resisting and sometimes embracing, from the very first episode of the series.

The call to adventure comes to Buffy as she sits on the steps of Hemery High School in Los Angeles. Suddenly, a man in a dark suit approaches, stands before her—arms perfectly perpendicular to the sidewalk, body straight as a rod—and announces: "Buffy Summers?... I need to speak to you.... There isn't much time. You must come with me. Your destiny awaits" (Whedon, 1998, "Becoming, Part I"). Buffy is clearly "weirded out" and with a half-licked lollipop in one hand, replies, "I don't have a destiny. I'm destiny-free, really." But this is not good enough for the man:

> Merrick: Yes, you have. You are the Chosen One. You alone can stop them.
> Buffy: Who?
> Merrick: The vampires.
> Buffy: Huh?

Buffy's confusion is too quickly resolved, as she soon finds herself clumsily wielding a wooden stake in a late-night training session in a local cemetery. Her Watcher, Merrick, paces between headstones, stoically calling out commands as Buffy fights her first vampire. Later, at home, her mother Joyce scolds her for being late and not calling; Joyce accuses her of being with her supposed ex-boyfriend, Tyler, against the wishes of both her parents. Later, in the bathroom, Buffy washes her hands and face. Her parents' arguing aurally frames her as she stares into the mirror, tears pooling and then streaming down her face.

Soon after Buffy's call to Slayerhood in Los Angeles, she is expelled from school for burning down the gym on prom night. Of course, she cannot explain to Hemery High's principal or her parents that the gym was full of vampires at the time and, thus, required torching. Buffy's expulsion seems to catalyze the disintegration of her parents' relationship, so after Joyce and Hank divorce, mother and daughter move to Sunnydale, both hoping to make a fresh start. Buffy particularly wishes to leave her calling behind. But when she stops by the library to borrow textbooks for her first day, Rupert Giles, the new librarian, realizes she is the Slayer he has been sent to train and offers her ancient texts about demons and vampires instead. Far from being pleased at his offerings, Buffy bolts as Giles turns to gather another dusty volume.

Later, when Buffy returns to give him a firm resignation, Giles counters with the same speech Merrick gave her the year before: "Into every generation..." She knows the myth, she exclaims, but insists she is retired, a declaration Giles cannot fathom. Why would a Slayer—*the* Slayer—reject her destiny? In the process of debating with Giles over relinquishing her title, Buffy inadvertently reveals herself to Xander, a very new acquaintance, who overhears their disagreement while looking for a library book. Before long, Xander lets his childhood companion Willow in on the secret. As the two develop a friendship with Buffy, the loveable but goofy Xander and witty but nerdy Willow soon find themselves in the middle of dangerous, supernatural skirmishes with evil. Once they accept that demons and vampires actually exist, though, both find a reason to laugh in the faces of their hipper and cooler—even if dumber, that is, ignorant—classmates; they feel needed, at least most of the time. While Buffy seems to be relieved she can share the burden of her calling with friends, she also worries about their safety, an added responsibility that surfaces in conversation often throughout all seven seasons, especially when the group faces an extraordinary crisis. As early as "The Harvest" (Whedon & Kretschmer, 1997), the first season's second episode, Buffy asserts her singularity. When Xander asks what plan "we" have for fighting the day's Big Bad, Buffy declares, "There's no 'we,' okay? I'm the Slayer, and you're not." Always already self-deprecating, Xander concedes to Buffy (for the time being). Willow, however, resists—even if gently—and Giles breaks Watcher Council guidelines by suggesting Willow be involved. The very next episode, "The Witch" (Reston & Cragg, 1997), repeats this kind of negotiating as the three begin to define their relationship with and to each other:

> Buffy: I need to get the skinny on Amber. Find out if she's had any colorful episodes before.
> Willow: That means hacking illegally into the school's computer system. At last, something *I* can do!
> Xander: I'll ask around about her.
> Buffy: You guys don't have to get involved.
> Xander: What d'ya mean? We're a team! Aren't we a team?
> Willow: Yeah! You're the Slayer, and we're, like, the Slayerettes!

These two early examples typify many conversations that attempt to define who is capable of doing what, who is responsible for whom, and who is in charge. The point of such repeated dialoguing (and sometimes arguing) is that the definitions Buffy, Willow, Xander (and later, key characters such as Angel, Cordelia, Oz, Spike, Faith, Anya, Riley, Tara, and Dawn) seek nearly always elude them. As a result, they perpetually try to define the who, what, where, when, why, and how of being the Slayer and being the Slayerettes.

Significantly, the Big Bad of the first season is an ancient vampire called the Master. This naming seems deliberate on Whedon's part, as he defines his *Buffy* endeavor as pointedly feminist. The Master, an "old one," stands in for structures of oppression, whether they oppress women such as Buffy or school clowns like Xander. That Buffy faces her fear of the Master, dies at his hand, is resuscitated/resurrected by Xander, and finally destroys her foe marks the beginning of the revision of the quest story. But even this revision is a journey in itself, an unraveling of strands that can then be refashioned for a new and better purpose.

## The Unraveling of the Myth

Throughout the twelve episodes of the first season, the Slayerettes develop into the Scooby Gang, slowly discovering and honing their skills as they increasingly assert their autonomy. This development of what are typically supporting but peripheral characters contrasts with the many examples of quest-driven stories. For example, *Stars Wars* is ultimately all about Luke, *The Matrix* about Neo, *Harry Potter* about Harry. In addition, whereas Buffy, again, slips back and forth between supporting her friends' growth and stifling it, between inviting their involvement and forbidding it, traditional heroes have no qualms about their singular role as savior. Though Buffy continues to vacillate up to the final episodes of the series, Willow and Xander arguably break away from conventional expectations for a hero's companions—voice of reason/pack animal (Frodo's Samwise), cheerleader (Neo's Trinity), spirit guide (Luke's Obi-Wan Kenobi)—in the first episode of the second season.

"When She Was Bad" (Whedon, 1997a) begins with Buffy's return from summer vacation in Los Angeles. Willow and Xander celebrate her arrival; the celebration, though, is short lived. During her initial encounter with Cordelia, the high school "Queen C" and recent unwilling addition to the Slayerettes, Buffy demands rather than requests Cordy's confidentiality with an insult: "You don't tell anyone that I'm the Slayer, and I won't tell anyone that you're a moron." Xander's dislike for Cordy and crush on Buffy prevent him from hearing the bitterness in Buffy's voice. However, the bitterness is not lost on Willow, who later notes that Buffy is different; she has never been cruel. Before either Willow or Xander, the audience understands that Buffy is haunted by dreams about the Master. When Buffy discovers that the Master's remaining servants plan to revive him, she comes undone bit by bit; her fear drives her to act completely out of character. Later at the Bronze, the city's only teen club and usual haunt of the Slayerettes, Buffy blows off Angel (her unattainable love interest whom she has discovered is a vampire with a soul) when he tries to warn her about the Master's minions, ignores Willow's feelings for Xander, and capitalizes on Xander's crush on her. Even her clothing is atypical: her short and slinky dress hugs her maturing body, a body she clearly and deliberately uses to entice Xander onto the dance floor. After seductively dancing with and around Xander, nearly kissing him, and leaving him alone, speechless and a little disgusted in the middle of the swaying crowd of couples, she exits the Bronze. Willow looks devastated by Buffy's obvious betrayal, and Angel appears at once jealous, angry, and disappointed. Arguably, it is at this violating moment that Willow and Xander cease to be Slayerettes, Buffy's sidekicks, and move toward more collective roles typified by a different title: Scooby Gang.[7]

If Buffy's disregard for her friends' feelings does not alter their roles in the group, her eventual self-righteousness and carelessness do. Later in the episode, Giles, Willow, and Xander try to figure out exactly what the Master's followers are doing. Buffy, however, thinks she has everything under control and needs no help or distractions from her friends. The scene is worth quoting at length to illustrate Buffy's arrogance and the emerging independence of Willow and Xander that will lead to their establishment as heroes equal to Buffy (Whedon, 1997a):

> Giles: Alright, alright, I-I-I've got something. It's Latin, so bear with me. Uh, to revive the vampire they need his bones, uh...w-which they have, and, um, the blood...this is very unclear, of the closest person...uh, someone connected to the vampire.
> Buffy: That'd be me.
> Giles: Perhaps.
> Buffy: We were close. We killed each other. It really promotes togetherness.
> Xander: Well, is there anything on when the ceremony might take...[A rock crashes through a window in the library.]...Whoa! [The rock is wrapped with a note and a bracelet.]
> Buffy: This is Cordelia's. "Come to the Bronze before it opens, or we make her a meal."
> Xander: They're gonna cook her dinner? I'll pretend I didn't say that.
> Willow: What do we do?
> Buffy: I go to the Bronze and save the day.
> Xander: I don't like this.
> Giles: Nor I!
> Buffy: Yeah? Well, you guys aren't going.
> Willow: What do you mean?
> Buffy: I can't do it anymore. I can't look after the three of you guys while I'm fighting.
> Willow: Well, what about the rest of the note?
> Buffy: What rest of the note?
> Willow: The part that says, "P.S. This is a trap"?
> Giles: You'll be playing straight into their hands.
> Buffy: I can handle this.
> Willow: Stop saying that! God, what's wrong with you?
> Xander: Cordelia may be dead.
> Buffy: This is my fight.

Her fight turns into the "trap" Willow predicted, and too quickly Buffy realizes that she has put the very people she thought she was saving in mortal danger. She rushes back to the library where she finds Xander beaten and bloody. In a decidedly nondeprecating moment, Xander declares that if Buffy had worked with the group, she could have stopped what is happening. Because Buffy pushed the Master through the library's skylight, not she but Giles, Willow, Cordelia, and Ms. Calendar (Sunnydale High's computer teacher) were the closest to him when he died. Xander abandons his usual humor and goofiness to make sure Buffy understands him: "If they hurt Willow, I'll kill you." As the episode climaxes and then comes to a close, only vampires meet their demise, and Buffy deals with her phobia of the Master by pulverizing his bones with a sledgehammer. The Scoobies forgive Buffy, and all is right with the world—until the next apocalypse, that is. The Sunnydale world, though, is inching toward change. Conscious or not of this change, the three teens continue on a revisionary path that redefines what it means to be a hero and aims to alter the world as they know it completely and forever.

## Toward a New Myth

The two-part, third season finale "Graduation Day" (Whedon, 1999a & 1999b) and, even more so, the penultimate, fourth season episode "Primeval" (Fury & Contner, 2000) illustrate the growing equitability of the characters and foreshadow the death and rebirth of the Slayer myth.

The Big Bad of season three is Sunnydale's Mayor Wilkins; his Ascension dominates the Scoobies' attention. Still, because it is their senior year, Buffy and friends must deal with the impending future of "real life"—SAT exams, college applications, and career choices. If these normal pressures are not enough, it soon becomes obvious from their research of ancient texts that, at the Ascension, Wilkins will become *pure* demon, one more massive in size and strength than any they have encountered before. Knowing she cannot fight the demon alone, Buffy devises a plan that will involve not only the Scoobies but also the entire senior class. More than ever, Buffy shares her responsibility as the Chosen One. It is during this season that we hear her say "Looks like a job for Wiccan girl" (Fury & Contner, 1999, "Choices") and "I'm going to need every single one of you on board, especially you Xander. You're sort of the key figure here" (Whedon, 1999b, "Graduation Day, Part II"). It is during this season that Buffy fires the Watchers' Council when they refuse to help her save Angel, breaking "laws that have existed longer than civilization" (Whedon, 1999a, "Graduation Day, Part I"). It is during this season's finale that Sunnydale High's pomp and circumstanced senior class is led by Xander,

Willow, Oz (Willow's boyfriend), and Angel in waves of attacks on the giant snake into which the mayor has ascended. In the end, Sunnydale High lies in ruins, mere rubble after the explosion necessary to kill the demon. Buffy, Xander, Cordelia, Willow, and Oz gather at the edge of campus. It is Oz who directs everyone to pause: "Guys, take a moment to deal with this: we survived" (Whedon, 1999b, "Graduation Day, Part II"). In this moment of reflection about the fight *and* four years of high school, there is no sense that Buffy owns the victory; rather, the circle of friends does.

Similar to the communal triumph of season three is the climax and resolution of the fourth season's story arc in "Primeval" (Fury & Contner, 2000). Before they can work together as a community of heroes, though, the arc's villain tries them in the fire. When Adam, an amalgam of monstrous and technological parts assembled by a covert military operation, goes awry, he quickly finds that the Slayer will not abide him and that the only way to destroy the Slayer is to lure her away from her friends. As Spike tells Adam, "You want her evening the odds in a fight, you don't want the Slayerettes mucking about" (Petrie & Grossman, 2000, "The Yoko Factor"). Instead of simply killing Xander, Willow, and the others, Adam uses Spike to stir up dissent among them so that, divided from her friends, the Slayer can easily be conquered. What Spike sets in motion leaves the group exactly as Adam wants them: disconnected. Buffy reverts to her old excuses, claiming that Adam is too dangerous for Willow and Xander to help. Willow calls Buffy out. Xander complains that he gets left out and underestimated because he has not been in college with them. In addition, Giles, drunk throughout the scene, whines about being unemployed and underutilized. Buffy puts an end to the spat by asserting her authority as the Slayer:

> Buffy: Enough! All I know is you want to help, right? Be part of the team?...No! No, you said you wanted to go. So let's go. All of us. We'll walk into that cave with you two attacking me and the funny drunk drooling on my shoe. Hey! Hey, maybe that's the secret way of killing Adam?!...Is that it? Is that how you can help? You're not answering me. How can you possibly help? So. I guess I'm starting to understand why there's no ancient prophecy about a Chosen One...and her friends (Petrie & Grossman, 2000, "The Yoko Factor").

When the Scoobies figure out what has happened, that Adam deliberately divided them in order to conquer them, they make up and rally. It is Xander who unwittingly devises the plan that will save the day:

"So...all we need is combo-Buffy—her with Slayer strength, Giles' multi-lingual know-how, and Willow's witchy power." Giles discovers that Xander is exactly right; their strategy will be the opposite of Adam's. Instead of divide and conquer, they will unite and stand.

Just before the final confrontation, the four gather to begin the ritual that will empower Buffy to face Adam. As Wilcox (2005) described, "Willow uses Tarot-like, large, named picture cards to identify each of the four main friends: Spiritus, spirit, for Willow...; Animus, Heart, for Xander; Sophus, Mind, for Giles; and Manus, the Hand, for Buffy" (p. 165). The names and pictures seem to correspond perfectly to the strengths of each Scooby, yet spirit, heart, mind, and hand also, together, just as easily compose one person, noted Wilcox. Here, then, is the perfect example of the equitability of the hero. Together, and calling upon the spirit of Sineya, the first Slayer, Buffy and her hero-friends defeat the monster. The narrow path of the lone hero widens to make room for a band of heroes who possess and hone their own skills while acting with the Slayer as a single force. At this point, the Scoobies have taken a sharper turn toward *Buffy*'s last season, when Buffy will finally not only know but also understand what they all must do to revise the Slayer myth.

## The Myth Revised

As much as she vacillates throughout the seasons between embracing and rejecting her identity as the Slayer, Buffy never wholly accepts the myth imposed upon her or the title "Chosen One" that comes with it. This rejection is what makes Buffy and *Buffy* different. Though Harry is flanked by Hermione and Ron, he faces his nemesis alone. Though Luke has a twin sister nearly as skilled as he is in combat, the battle that counts the most is between Luke and his father. Even Frodo, accompanied by Samwise throughout all but the final moments of the journey, finally faces the fires of Mordor alone. These heroes, at some point in their quests, embrace their destiny, assume the titles they have been assigned. Buffy, though, misbehaves. Though Playden (2001) argued that being the Slayer "is fundamental to [Buffy's] being" (p. 126), Buffy's title still does not completely reassure her of who she is. Playden cited Kendra (the Slayer activated after Buffy's first death in season one) in "What's My Line, Part II" (Noxon & Semel, 1997) who thanks Buffy for teaching her that slaying is not just a job but a way of being, an identity—"It's who you are." But the audience does not know whether Buffy shares this sentiment, as she simply smiles in response rather than speaking. In fact, it is not until she and her friends bend the myth to their own lives and context that Buffy, along with her friends, seems to discover and claim her true self. When Spike calls Buffy "the one" and she responds "I don't want to be the one" (Kirshner, Noxon, & Contner, 2003, "Touched"), her statement can be read as a refusal of the central and solitary role she is supposed to play according to the ancient Slayer myth: one girl in all the world, a girl whose power, we learn in the final season, comes from a long line of women, the first of whom was forced to accept supernatural strength conjured by a group of male, tribal leaders thousands of years before. When Buffy herself is psychically transported to this same circle of men, she refuses the power they offer her so that she can defeat the First. Instead, the men give Buffy a vision of the impending battle with the First's unlimited armies. Now Giles' earlier words weigh heavier than they ever have:

> The First is unlike anything we've faced before. I mean, there's evil and then there's the thing that created evil, the source.... That's what it is. It has eternities to act, endless resources. How to defeat it...I-I-...honestly, I don't know. But we have to find a way. If the Slayer line is eliminated, then the Hellmouth has no guardian. The balance is destroyed. I'm afraid it falls to you, Buffy. Sorry. I mean, we'll do what we can, but you're the only one who has the strength to protect...the world against what's coming (Noxon & Petrie, 2002, "Bring on the Night").

The vision forces Buffy to question her choice. Should she have taken the power, even if it would have been a violation of both her body and soul? Though she is not sure of her decision, her rejection of the men's mystical power becomes yet another step in the direction of transforming their legacy, of reshaping the myth.

On her return from her psychic journey, Buffy's energy and determination to fight wane. It is now that her fear turns her into a seemingly power-hungry, bossy military leader, and it is now that her friends and family depose her. Her short time away from the troops and Spike's reassurance that she is more than the Slayer recharge Buffy. Finally, she "gets it": she cannot save the world alone; one hero, even one with supernatural power, is not sufficient. As many scholars have noted and Wilcox (2005) articulated, "Buffy, despite the voice-over which intones that she is the only one, actually repudiates patriarchal succession and the role of the lonely hero in favor of communal effort" (p. 76; see also Miller, 2003; Playden, 2001; Rose,

2002; South, 2001). Here is where the foreshadowing of previous season finales—especially the ones described earlier—climaxes. Buffy realizes that "communal effort" does not even mean Scooby Gang effort. The First and its armies would annihilate the Scoobies with little exertion. No, the effort will require much more than spirit, heart, mind, and hand.

In the scenes leading up to the final moments of the series, we watch as the Slayer myth, a version of the traditionally understood monomyth, is given a new form. Buffy walks everyone—her friends, her sister, Spike, Faith (the Slayer triggered by Kendra's death), and the handful of potentials who have gathered in Sunnydale—through the strategy. They will fight a physical battle, but they will also, through a ritual performed by Willow, empower all potential Slayers all at once, breaking the ancient decree that reserved power for one woman at a time. This plan is in contrast to the way many orthodox heroes—Superman, Black Alice, Highlander—borrow or even commandeer the power of others to emerge as lone *super*heroes. Buffy's many experiences along her shared journey with her friends have taught her that more strength resides in the sharing rather than the taking of power. Ultimately, then, the plan Buffy proposes works, and when all is said and done, the Hellmouth collapses, leaving Sunnydale a mere desert crater (Whedon, 2003, "Chosen"):

> Faith: Looks like the Hellmouth is officially closed for business.
> Giles: There's another one in Cleveland. Not to spoil the moment.
> Xander: We saved the world.
> Willow: We *changed* the world. I can feel them, Buffy. All over. Slayers are awakening everywhere.
> Dawn: We'll have to find them.
> Willow: We will.... Yeah, the First is scrunched, so...what do you think we should do, Buffy?
> Faith: Yeah, you're not the one and only Chosen anymore. Just gotta live like a person. How's that feel?
> Dawn: Yeah, Buffy. What are we gonna do now?

The scene speaks for itself. Xander notes that they saved the world—again; in fact, saving the world is not unusual for the Scoobies. Willow corrects him; more than saving it, they have changed the world, and by extension, they have "reformed" the quest story. Buffy is no longer the only one who is chosen, and that fact did not transpire because of her: Buffy did not change the world; everyone involved did. In *Buffy*, the sharing of battle, of power, of triumph demonstrates the equitability of the hero.

## Discussion: All Is in the Group

*We* saved the world. I say we party.

—Buffy (Whedon, 1997b, "Prophecy Girl," emphasis added)

The equitability of the hero is significant for a new millennium of viewers exactly because it revises traditional understandings of the monomyth. Pearson and Pope (1981) noted that "our understanding of the basic spiritual and psychological archetype of human life has been limited...by the assumption that the hero and central character of the myth is male" (p. 4). Not only is the hero male but also white, upper class, and always already powerful in some way—politically, financially, socially. Even when attempts are made to avoid this typical construction, Pearson and Pope insisted that the "patriarchal habit" takes over; too quickly the pronoun *he* dominates. Even Campbell (see also Norman, 1969), who assumes the hero may be male or female, reverts to using the male pronoun throughout his classic text. Several notable narratives, though, attempt to break the patriarchal habit by humanizing, instead of gendering, the hero. For instance, "one of the major themes in Bertolt Brecht's German drama *Mother Courage and Her Children* is that people who are not recognized by the society as noble or heroic often slay dragons out of necessity" (Pearson & Pope, 1981, p. 5). Xander is an ideal example. When Dawn thinks she has been chosen as a would-be Slayer in "Potential," she finally feels valuable, her own person instead of the Slayer's little sister. But all too quickly, Dawn and the whole group discover that she was mistaken. What Xander tells Dawn later, when they are alone together, illustrates how the most ordinary person is still a hero (Kirshner & Contner, 2003, "Potential"):

Dawn: What's up?
Xander: Aw, I'm just thinking about the girls. It's a harsh gig, being a potential. Just being picked out of a crowd, danger, destiny...plus, if you act now, death.
Dawn: They can handle it.
Xander: Yeah. They're special, no doubt. The amazing thing is, not one of them will ever know, not even Buffy.
Dawn: Know what?
Xander: How much harder it is for the rest of us.
Dawn: No way. They've got...
Xander: Seven years, Dawn. Working with the Slayer. Seeing my friends get more and more powerful. A witch. A demon.... Powerful. All of them. And I'm the guy who fixes the windows.

> Dawn: Well, you had that sexy army training for a while, a-and the windows really did need fixing.
> Xander: I saw what you did last night.
> Dawn: Yeah, I...guess I kinda lost my head when I thought I was the Slayer.
> Xander: You thought you were all special. Miss Sunnydale 2003. And the minute you found out you weren't, you handed the crown to Amanda without a moment's pause. You gave her your power.
> Dawn: The power wasn't mine.
> Xander: They'll never know how tough it is, Dawnie, to be the one who isn't chosen. To live so near to the spotlight and never step in it. But I know. I see more than anybody realizes because nobody's watching me. I saw you last night. I see you working here today. You're not special. You're extraordinary.
> Dawn: Maybe that's your power.
> Xander: What?
> Dawn: Seeing. Knowing.
> Xander: Maybe it is. Maybe I should get a cape.
> Dawn: Cape is good.

Xander is never comfortable believing he could be in the "spotlight," yet it is clear from his conversation with Dawn that Whedon (with Kirshner and Contner) intends for both Xander and Dawn to be understood as heroes, even superheroes: "Cape is good." However private or off center stage, the moment *is* Xander's and Dawn's spotlight.

Campbell was right: we keep telling quest stories across time and cultures. And while they may be similar enough from a distance to warrant the construction of a monomyth, there is danger in reducing any person's or culture's story or stories to one. The monomyth is a telescopic, "big picture" tale, not a microscopic, detailed one. Variations are abundant and those variations matter. The shades of difference in the myths we share from generation to generation, from society to society, matter because those shades highlight what a person, a people, a nation values. Joss Whedon values equality and, maybe even more importantly, family. Not the family of American tradition—mother, father, daughter, son, and a dog or cat. He believes instead in the chosen family (Battis, 2005), one in which patriarchy and its effects are shattered: lone, male heroes and companions who are simply sidekicks. In the place of patriarchy, equality and *its* effects are celebrated: the collective hero, the equitability of journeyers. All share the boon; all are empowered to serve the community. All have, if not supernatural powers, (extra)ordinary skills, talents, and gifts.

Too many scholars seem to overlook the final pages of *The Hero with a Thousand Faces*, reducing the monomyth to what we have described throughout this argument and ourselves attributed to Campbell: the journey of a solitary, privileged male. But this standard construction of the Campbellian hero is not as set in stone as we have led ourselves to believe. According to Campbell himself (1949), "The problem of [humankind] today...is precisely the opposite to that of [people] in the comparatively stable periods of those great co-ordinating mythologies which now are known as lies. Then all meaning was in the group...none in the self-expressive individual; today no meaning is in the group—none in the world: all is in the individual" (p. 388). The world has evolved; much of what was once myth or mystery has been analyzed by scientists or theorists, proven "true" or "false." So the challenge, claimed Campbell, is how to become fully human through the circumstances we find ourselves in right now. This challenge is exactly what *Buffy the Vampire Slayer* attempts to meet, and its answer lies in a turn back to the collective, one beginning to be echoed throughout popular culture through new series such as NBC's *Heroes*, through fan fiction, and through multiple episodic authors. Willow was right: "We changed the world" (Whedon, 2003, "Chosen"). More and more, meaning is being found not in the individual but in the group.

## Implications: Where to Go from Here

I mean, I know every Slayer comes with an expiration mark on the package, but I want mine to be a long time from now. Like a Cheeto...

—Buffy (Petrie & Marck, 2000, "Fool for Love")

Throughout this chapter, we have established that *Buffy the Vampire Slayer* "misbehaves" as it (re)enacts the monomyth. Specifically, we maintain that Buffy's companions, the Scoobies, take a much more active role in the quest than is typical. Willow, Xander, Giles, and many others become heroes in their own right. Their knowledge and abilities as well as their loyalty make them not only indispensable to Buffy's journey but also integral to the construction of a new myth, one that changes and benefits the world by distributing power equally. In short, the sidekicking Slayerettes grow into the center-staged Scoobies, partners with Buffy in the fight against evil. We have further argued that throughout the series Buffy struggles with her identity as the Slayer. Early on, she establishes the general hierarchy of the hero's journey, but eventually Buffy shares her strength with all potential Slayers, inviting them into the equitability of heroic deeds already held in common with her friends, her family. This variation on a theme revises myth for modern times. Beyond what we have argued here, our analysis suggests several implications.

First, our reading reveals that *Buffy*'s revisions to the quest are reflective of an evolving culture that plays with and often breaks the mold of heroic archetypes. While male, loner hero types such as those played by Arnold Schwarzenegger (Conan), Sean Connery et al. (James Bond), and Vin Diesel (Riddick) still abound, *Buffy* is one among a growing number of examples in which traditionally unexpected and disparate heroes from unusual places are becoming prominent in American culture. Buffy, of course, enjoys notable company—Xena: Warrior Princess, La Femme Nikita, Ripley, and Sidney Bristow, among others.[8] Women are not alone in their rise to hero status, though. With the fall 2006 premiere of NBC's *Heroes*, we see that very (extra)ordinary people around the world will lead humanity, via genetic evolution, into a new era (the allusion to *X-Men* is obvious, though the title itself—*Heroes*—does away with both the *men* in *X-Men* and the negative connotations of the term *mutant*). Even the largely scripted realm of "reality" TV is a breeding ground for heroes who are of all genders, races, and classes. Television that imitates or (re)produces "real life" may eventually drive audiences and scholars to question how praiseworthy or even remarkable is the whole notion of a hero. Is our most recent Superman correct when he insists that the world longs for a savior-hero, or is Lois Lane correct when she insists that we do not?

Second, we demonstrate the need for further research into how the hero's journey, as well as other mythic ideals, is currently being recast. While the quest story still serves as a standard framework for many films and television series, *Buffy* shows how some of its characteristics have changed or been crafted to fit present circumstances. Surely, *Buffy* is not the only series within which the monomyth has been revised or entirely altered. Even as this collection contributes to the conversation of modern mythologies, we hope to see more analyses, especially of television, that use the framework of the quest to critique culture, and culture to critique the quest. Doing so will perhaps promote a better understanding of how humans use story and myth in this new millennium and, in turn, lead to deeper insight into what it means to be human.

Finally, *Buffy* demonstrates that television is capable of hosting heroes who can be self-reflective, (r)evolutionary, inclusive, and communal. Jewett and Lawrence (1977, 2002) argued that Hollywood uses the quest story as a basic template for the creation of its heroes. They also maintained that American heroes, as depicted in the movies, experience no character growth; they do not gain understanding or wisdom that they wish to share. This lack of growth and refusal to share knowledge create a culture in which heroes believe they can become solitary saviors rather than participants in a community. Our examination of *Buffy* and the Scooby Gang, however, reveals a reformation of that kind of hero. Buffy struggles with her identity as the Slayer and comes to rely more and more upon her friends. As Buffy takes this journey, her friends do as well; they give just as much as they receive. Essentially, Buffy creates a network of relationships that prove vital in order to deal with the dangers of the outside world. This network of relationships was absent from Jewett and Lawrence's work. Our case study of *Buffy* suggests that Hollywood, as well as American culture at large, can offer heroes who are ordinary, diverse, and human, ones whom audiences can identify with and emulate precisely because they have the capacity to change and recognize their need of others. Thus, Jewett and Lawrence's concern may become misplaced in the twenty-first-century.

*Buffy the Vampire Slayer* openly subverted a number of mythical archetypes over its seven-season run and continues to do so as new converts turn to the series in syndication and on DVD. It provided entertainment and education for millions and has made us amend how we think about the typical heroic image within mythology. One can only hope that the future of television—or any creative medium—is marked by *Buffy*'s (re)visionary and transformational cultural work, which, much like the Cheeto, refuses to expire.

## Notes

1 In their analysis of the *Buffy* video game released in 2002 for *X-Box*, Labre and Duke (2004) found that "the type of heroism portrayed in the *Buffy* video game is characterized by the Warrior archetype, the male shadow, and an ethic of justice rather than care" (p. 152). This characterization complicates a more progressive and positive depiction of the female just warrior (Early, 2001) and is in stark contrast to the series' portrayal of heroism.

2 Buffy is not alone in being an unorthodox heroic figure. J.R.R. Tolkien's Frodo, for instance, does not embody the stereotypical archetype. Nevertheless, rather than simply bending the stereotype, Buffy redefines what it means to be a hero.

3 The Slayer myth dictates that when the Chosen One dies, a new Slayer is called; thus, a pool of "potentials" is always being trained for the possible position opening.

4 It is significant to note that Nimmo and Combs' naming of the myth is gendered: "*master* myth" (emphasis added). Pearson and Pope (1981) would call this naming evidence of the "patriarchal habit" to view the hero as always already male.

5 In "Family" (Whedon, 2000a), the father, brother, and sister of Willow's girlfriend Tara come to Sunnydale to take Tara home so that she can assume her "rightful" subservient role in the home. Buffy stands up to the father, exclaiming that he will have to go through her first in order to take Tara against her will; Dawn seconds, and the rest of the Scooby Gang follows suit. Stunned by the group's audacity, the man counters, "You people have no right to interfere with Tara's affairs. We're her blood kin. Who the hell are you?" Buffy calmly and firmly replies, "We're family."

6 Certainly others have challenged the image of the invulnerable protagonist (Donaldson's Thomas Covenant and Tolkien's Frodo, for example). Nevertheless, few have gone so far and have been so deliberate in doing so as Whedon's Buffy.

7 Willow says as much in "Fear, Itself" (Fury & Gates, 1999). When the Scoobies attend a Halloween party at a local frat house, something mystical begins to happen. The doors disappear, and the group gets lost inside of the house. When Buffy suggests that her friends are getting in the way of her duties as the Slayer, Willow does not let her get away with labeling them a hindrance: "Being the Slayer doesn't automatically make you the boss.... I'm not your sidekick."

8 For a discussion that complicates the notion of Xena, Nikita, and Buffy as ideal feminist heroes, see Magoulick (2006).

# References

Battis, J. (2005). *Blood relations: Chosen families in* Buffy the Vampire Slayer *and* Angel. Jefferson, NC: McFarland & Company.

Bowman, L. (2002). *Buffy the Vampire Slayer:* The Greek hero revisited. Retrieved May 12, 2006, from http:// web.uvic.ca/!lbowman/buff/buffythehero.html.

Buttsworth, S. (2002). "Bite me": Buffy and the penetration of the gendered warrior-hero. *Continuum: Journal of Media & Cultural Studies, 16* (2), 185–199.

Campbell, J. (1949). *The hero with a thousand faces.* New York: MFJ Books.

Cauthen, B. (2004). Covenant and continuity: Ethno symbolism and the myth of divine election. *Nations and Nationalism, 10* (1/2), 19–33.

Dorsey, L. G. (1995). The frontier myth in presidential rhetoric: Theodore Roosevelt's campaign for conservation. *Western Journal of Communication, 59* (1), 1–19.

Early, F. H. (2001). Staking her claim: *Buffy the Vampire Slayer* as transgressive woman warrior. *Journal of Popular Culture, 35* (3), 11–27.

Edwards, J. A. (2000). The demonic redeemer figure in political myth: A case study of Vladimir Zhirinovsky. *Journal of the Wisconsin Communication Association, 31*, 17–32.

Fulmer, H. W. (1986). Mythic imagery and Irish nationalism: Henry Grattan against Union, 1800. *Western Journal of Speech Communication, 50*, 144–157.

Fury, D. (Writer) & Contner, J. A. (Director). (1999). Choices. [Television series episode]. In J. Whedon (Creator), *Buffy the Vampire Slayer*.

Fury, D. (Writer) & Contner, J. A. (Director). (2000). Primeval. [Television series episode]. In J. Whedon (Creator), *Buffy the Vampire Slayer*.

Fury, D. (Writer) & Gates, T. (Director). (1999). Fear, itself. [Television series episode]. In J. Whedon (Creator), *Buffy the Vampire Slayer*.

Greenwalt, D. (Writer) & Kretschmer, J. T. (Director). (1997). School hard. [Television series episode]. In J. Whedon (Creator), *Buffy the Vampire Slayer*.

Jewett, R. & Lawrence, J. S. (1977). *The American monomyth*. Grand Rapids, MI: Eerdmans.

Jewett, R. & Lawrence, J. S. (2002). *The myth of the American superhero*. Grand Rapids, MI: Eerdmans.

Kirshner, R. R. (Writer) & Contner, J. A. (Director). (2003). Potential. [Television series episode]. In J. Whedon (Creator). *Buffy the Vampire Slayer*.

Kirshner, R. R. (Writer), Noxon, M., & Contner, J. A. (Directors). (2003). Touched. [Television series episode]. In J. Whedon (Creator). *Buffy the Vampire Slayer*.

Kluver, A. R. (1997). Political identity and national myth: Toward an understanding of political legitimacy. In A. Gonzalez & D. V. Tanno (Eds.), *Politics, communication, and culture* (pp. 48–75). Thousand Oaks, CA: Sage.

Labre, M. P., & Duke, L. (2004). "Nothing like a brisk walk and a spot of demon slaughter to make a girl's night": The construction of the female hero in the *Buffy* video game. *Journal of Communication Inquiry, 28* (2), 138–156.

Magoulick, M. (2006). Frustrating female heroism: Mixed messages in *Xena, Nikita,* and *Buffy*. *Journal of Popular Culture, 39* (5), 729–755.

Miller, J. P. (2003). "The I in team": Buffy and feminist ethics. In I. William (Series Ed.) & J. B. South (Vol. Ed.), Buffy the Vampire Slayer *and philosophy: Vol. 4. Fear and trembling in Sunnydale* (pp. 35–48). Chicago, IL: Open Court.

Nimmo, D. & Combs, J. E. (1980). *Subliminal politics: Myths and mythmakers in America*. Englewood Cliffs, NJ: Prentice-Hall.

Norman, D. (1969). *The hero: Myth/image/symbol*. New York: World Publishing.

Noxon, M. (Writer) & Petrie, D. (Director). (2002). Bring on the night. [Television series episode]. In J. Whedon (Creator), *Buffy the Vampire Slayer*.

Noxon, M. (Writer), & Semel, D. (Director). (1997). What's my line, part II [Television series episode]. In J. Whedon (Creator), *Buffy the Vampire Slayer*.

Owen, S. (1999). Vampires, postmodernity, and postfeminism: *Buffy the Vampire Slayer*. *Journal of Popular Film and Television, 27* (2), 24–31.

Pearson, C. & Pope, K. (1981). *The female hero in American and British literature*. New York: R. R. Bowker.

Petrie, D. (Writer) & Grossman, D. (Director). (2000). The Yoko factor. [Television series episode]. In J. Whedon (Creator), *Buffy the Vampire Slayer*.

Petrie, D. (Writer) & Marck, N. (Director). (2000). Fool for love. [Television series episode]. In J. Whedon (Creator), *Buffy the Vampire Slayer*.

Playden, Z. (2001). "What you are, what's to come": Feminisms, citizenships and the divine ethic. In R. Kaveney (Ed.), *Reading the vampire slayer: An unofficial critical companion to* Buffy *and* Angel (pp. 120–147). New York: Tauris Parke.

Reston, D. (Writer) & Cragg, S. (Director). (1997). The witch. [Television series episode]. In J. Whedon (Creator), *Buffy the Vampire Slayer.*
Rose, A. (2002). Of creatures and creators: *Buffy* does *Frankenstein.* In R. V. Wilcox & D. Lavery (Eds.), *Fighting the forces: What's at stake in* Buffy the Vampire Slayer (pp. 133–142). Lanham, MD: Rowman & Littlefield.
Rushing, J. H. & Frentz, T. S. (2005). The mythic perspective. In J. Kuypers (Ed.), *The art of rhetorical criticism* (pp. 241–269). New York: Allyn & Bacon.
South, J. B. (2001). "All torment, trouble, wonder, and amazement inhabits here": The vicissitudes of technology in *Buffy the Vampire Slayer. Journal of American and Comparative Cultures, 24* (1/2), 93–102.
Starr, J. B. (1973). *Ideology and culture.* New York: Harper & Row.
Whedon, J. (Writer & Director). (1997a). When she was bad. [Television series episode]. In J. Whedon (Creator), *Buffy the Vampire Slayer.*
Whedon, J. (Writer & Director). (1997b). Prophecy girl. [Television series episode]. In J. Whedon (Creator), *Buffy the Vampire Slayer.*
Whedon, J. (Writer & Director). (1998). Becoming, part I. [Television series episode]. In J. Whedon (Creator), *Buffy the Vampire Slayer.*
Whedon, J. (Writer & Director). (1999a). Graduation day, part I. [Television series episode]. In J. Whedon (Creator), *Buffy the Vampire Slayer.*
Whedon, J. (Writer & Director). (1999b). Graduation day, part II. [Television series episode]. In J. Whedon (Creator), *Buffy the Vampire Slayer.*
Whedon, J. (Writer & Director). (2000a). Family. [Television series episode]. In J. Whedon (Creator), *Buffy the Vampire Slayer.*
Whedon, J. (Writer & Director). (2000b). Restless. [Television series episode]. In J. Whedon (Creator), *Buffy the Vampire Slayer.*
Whedon, J. (Writer & Director). (2003). Chosen. [Television series episode]. In J. Whedon (Creator), *Buffy the Vampire Slayer.*
Whedon, J. (Writer) & Kretschmer, J. T. (Director). (1997). The harvest. [Television series episode]. In J. Whedon (Creator), *Buffy the Vampire Slayer.*
Wilcox, R. V. (2005). *Why* Buffy *matters: The art of* Buffy the Vampire Slayer. New York: I. B. Tauris.

•CHAPTER NINE•

# Makin' Whoopi: Race, Gender, and the Starship *Enterprise*

*Stephanie Kelley-Romano*[*]

For more than forty years *Star Trek* has been a worldwide phenomenon, producing four spin-off television series, ten feature films, a Saturday morning cartoon series, numerous conventions, Internet sites, magazines, encyclopedias, and action figures. The durability of these texts, evidenced by the frequency of reruns, speaks to the continuing resonance *Star Trek* has for audiences in the new millennium. *Star Trek* seems to be a quintessential American text—a text that articulates contemporary values regarding race and gender. Rhonda Wilcox (1996) notes the dominant themes of the show often "parallel contemporary socio-political issues" (p. 69).

Science fiction is generally acknowledged to intentionally challenge and critique social practices. For example, Sliverblatt (2007) notes that "the futuristic setting of the science fiction genre furnishes perspective on political and social issues and events" (p. 148). As a cultural commentary, *Star Trek: The Next Generation* (*TNG*) has been the focus of much scholarly attention.[1] Many of these studies seek to understand the "representational nature and emblematic status" of *Star Trek* (Harrison, Helford, Ono, & Sarah, 1996, p. 2, 1996, p. 2). Many critics recognize that current issues are often reflected in the dominant themes of the program. For example, Clyde

[*] Dr. Stephanie Kelley-Romano (M.A., Emerson College; Ph.D., University of Kansas) is an Associate Professor of Rhetoric at Bates College in Lewiston, Maine.

Wilcox (1992) examines the conservative nature of cultural change represented in *Star Trek*. As a transmitter of culture, issues of gender, race, class, and sexuality are common themes in both *TNG* and the resulting body of criticism. Casper and Moore (1995) attend to sex and reproduction on the starship *Enterprise*. Both Roberts (1999b) and Projansky (1996) analyze issues and controlling images surrounding the issue of intergalactic violation and rape—demonstrating *Star Trek* is an appropriate space to play out social and cultural issues. Similarly, Bernardi (1998) explores the ways in which "race is produced, performed, and perpetuated" on the *Enterprise* (p. 11). More specifically, VandeBerg (1996) argues the assimilation of the Klingon Worf results in the suppression of cultural difference, and Wilcox (1996) argues the android Data's extreme whiteness characterizes him as a racial "other." Each of these studies demonstrates the importance of understanding cultural representations as a way of transmitting stereotypes of minorities, and the perpetuation of positioning "others" in socially inferior spaces.

*Star Trek: The Next Generation* is particularly worthy of critical attention since it purports to represent the idealized future. However, as noted by Bernardi (1998) because there are no real space-time referents for extraterrestrial races, aliens can always be said to represent real-world people (p. 12). The dialectical tension that is present in science fiction texts symbolizes the difference between the imagination of writers/producers concerning the potentialities of the future and the degree to which they are mired in traditional myths and stereotypes surrounding race, class and gender. Analysis of *Star Trek*, and particularly the characters who personify these tensions, makes these spaces visible. This chapter pays particular attention to the myths portrayed and perpetuated through the character of Guinan on *TNG* in an attempt to understand the specific themes and messages spoken to, and about, black women.

The bodies of black women are commonly the stages on which cultural stereotypes are displayed. These stereotypes appeal to viewers because they evoke long-standing, explanatory mythic themes. Research has indicated that these "mediated portrayals of African American women influence judgements of African American women in social situations" by "priming" viewers to receive images in particular ways (Brown-Givens & Monahan, 2005, p. 87). In *TNG*, the *Enterprise* continues its tradition of having a female onboard the ship who can serve as the stereotypical representation of "black womanness." These negative stereotypes, notes Patricia Hill Collins (2000), "have been fundamental to Black women's oppression" (p. 72).[2] In the original 1960s series, Lieutenant Uhura, a communications officer played by Nichelle Nichols, was liberating in so far as she was the first black woman to serve an important role on the ship's bridge. However, with her micro-mini uniform and involvement in the first implied interracial kiss on television she was objectified and embodied the exotic "other."[3] Despite the oversimplified nature of her character, as a black actress with a permanent role on a major television drama she became a role model for many young girls in the United States. So began the struggle between lauding the fact a black woman could be an important part of a galaxy-hopping crew and condemning her as an overly sexualized token. Crediting Nichelle Nichols with inspiring her to become an actress, Whoopi Goldberg approached the creators *TNG* and requested a part on the show.

Goldberg was written in as Guinan, bartender at Ten-Forward, the ship's lounge. Although she appears in only twenty-eight episodes, Guinan is a particularly rich source for understanding representations of race and gender aboard the *Enterprise* in the twenty-fourth century. Roberts (1999a) notes that for those characters in whom race and gender intersect, "representation becomes even more complex and troubling" (p. 127). Historically, characterizations of black women in the media have "served to maintain systems of racial inequality and White domination" (Cox, 2005). In this chapter, I argue that the character Guinan, rather than boldly going where no one has gone before, is instead domesticated and confined to the traditional female role of caretaker. A close textual analysis of the twenty-eight episodes and one feature film (*Generations*) in which Guinan appears indicates her physical appearance as well as her roles as bartender, interpersonal advisor, and maternal figure reinforce stereotypical myths of black women and specifically evoke mythic conceptions of the "mammy" and the "jezebel." It is important that these myths, as transmitters of cultural values, are made transparent. Additionally, I argue that there is a clear duality between the public Guinan and the private Guinan and that each appeals to a different myth of black women. The public Guinan promotes values and behaviors commonly associated with the mammy, whereas the private Guinan provides writers/producers an opportunity to portray the sexually charged stereotype of the jezebel. This coupling is particularly troublesome because, at first glance, Guinan seems independent and influential; however, systematic analysis reveals her agency is restricted to a sphere defined by white male power. Her character is mythically significant in that many of the qualities associated with Guinan go unquestioned and seem natural for a black woman to exhibit.

Guinan is a racially and physically isolated character, and as such, what she represents becomes all the more significant. There are several general aspects of the character Guinan that distinguish her as an isolated individual. First, Guinan is one of the few surviving members of the El-Aurian race. The El-Aurians were attacked by the Borg, a collective whose skin tone turns pale gray on assimilation.[4] Typical of diasporic people, the El-Aurians were forced to scatter across the universe. Guinan managed to escape the Borg attack via a transporter ship rescued by the *Enterprise*. She was then allowed to remain onboard as hostess of Ten-Forward. Further serving to isolate her is the fact that she is the only non-Starfleet graduate in a position of authority/power on the *Enterprise*. As such, she is visibly marked in that she cannot wear the Starfleet uniform, and instead is costumed in bold, sometimes absurd garments. As a character, Guinan is within—but not a part of—*Enterprise* culture; in short, she is perfectly situated to continue the longstanding tradition of marking black women as different from the norm.

## The Mammy

The mammy is one of the oldest representations of black women and, as such, is one of the hardest to escape. Bogle (1997) notes "once the basic mythic types" evolved during slavery, the stereotypes could emerge "in various guises" (p. 17). Although she may have lost her bandanna and a few pounds, mammy is still the underlying essence of many black female characters on television. Collins (2000) notes that mammy is "central to intersecting oppressions of race, gender, sexuality and class" (p. 73). Although many scholars have argued that the stereotype of the mammy is based more in myth than in history, its origin is irrelevant to its impact on popular culture. Litwack (1982) recognizes that the perpetuation of happy, loyal, contented slaves "reflected white needs more than it did reality" (p. 19). Thurber (1992) quotes Parkhurst's popular description of mammy:

> Self-respecting, independent, loyal, forward, gentle, captious, affectionate, true, strong, just, warm-hearted, compassionate-hearted, fearless, popular,

> brave, good, pious, quick-witted, capable, thrifty, proud, regal, courageous, superior, skilful, tender, queenly, dignified, neat, quick, tender, competent, possessed with a temper, trustworthy, faithful, truthful, neither apish nor servile (Quoted in Thurber, 1992, p. 93)

Guinan, as a black female, working-class El-Aurelian stands at the point of this intersection.

The characterization of Guinan as mammy is immediately made apparent to viewers through her physical appearance. Traditionally, mammy was "big, fat, and cantankerous" (Bogle, 1997, p. 9). One of her most distinctive characteristics was the bandanna she wore on her head. Practically, bandannas served to keep a woman's hair up and out of the way while she worked simultaneously keeping her appearance neat. Bandannas also served to tie up and conceal part of what makes a woman unique and sensual—her hair. Like mammy, Guinan covers her hair—but rather than a bandanna she wears large, funky, hats. These hats are sometimes so large they flop around (see "The Child," "In Theory," "Rascals") and move her as far away from sensual and as close to silly as is possible on a dramatic series. Her head, unlike any other head on the *Enterprise,* is always covered. For example, "I Borg" opens with Guinan and Captain Picard fencing. Each is equipped with full-body padding and mesh face mask. Guinan, however, has wrapped her hair in silver fabric, continuing her practice of always covering her head and hair in public, thereby keeping that which makes her a sexual being hidden. Mammy Guinan always exhibits the proper amount of modesty.

Guinan's covering of her head and hair is particularly salient when contrasted with Deanna Troi, counsellor of the *Enterprise*. Troi, who as half-human half-Betazoid is white, is clearly the most sensualized character in the show. Roberts (1999a), in her examination of the sexual politics of *Star Trek*, notes that Troi's heritage and empathic abilities make her "the perfect mate" (p. 67). Her hair, which is never covered, marks her as a sensual being and conforms to traditional notions of femininity. Guinan, on the other hand, is not seen in public as a sexual being and instead is presented as asexual.

The mammy stereotype is also visually constructed through clothing. In popular portrayals, mammy's outfit was a modest one that consisted of a full-length, long-sleeved housedress. Over this, mammy wore an apron. Like her predecessor, Guinan wears brightly colored, heavy, loose-fitting robes, which disguise her size and add bulk to her frame. Deviation from standard Starfleet-issued uniforms serves as a visual cue and physical marker of difference. Her robes stand in stark contrast to the fitted uniforms worn by other individuals on the *Enterprise*. Interestingly, all of the major racially marked characters are distinguished by differences in uniform. Worf, Klingon chief of security, wears his Klingon warrior sash over his Starfleet uniform; Geordi La Forge, chief of engineering, wears a visor to protect his sensitive eyes. Likewise, uniform deviation highlights gender differences between crewmembers. Dr. Beverly Crusher, chief medical officer, wears long multicolored lab coats over her uniform; Deanna Troi is often seen wearing long flowing skirts and is always seen sporting generous cleavage. Troi's outfits (and she seems to be the only character who actually has outfits) although different from traditional uniforms, still are uniform in color and have piping that makes them uniform-like. Guinan's dress, however, looks nothing like a uniform. In fact, her gowns resemble "stereotypical African dress" and "thus help to code her story as one analogous to the 'diaspora' of Africans who were 'assimilated' and then 'regenerated' as slaves in the United States" (Hastie, 1996, p. 120). Likewise, Roberts (1999a) notes Guinan "reifies the sense of African-Americans as almost literally alien" (p. 128). Clearly then, Guinan's clothing serves to set her apart from the rest of the crew, mark her as "other," and associate her with stereotypical notions of the mammy.

In addition to evoking memories of mammy through clothing, the physical space Guinan occupies reinforces her characterization as mammy. Traditionally, mammy's domain was the kitchen. Reinforced by the cultural icon of Aunt Jemima, black women's active space has been, and continues to be, limited to the domestic arena and her perceived competence limited to the goodness of cleanliness and food. Therefore, mammies became maids; slaves became servants. Interesting then, is Guinan's positioning in Ten-Forward. In the twenty-fourth century technological advances have made cooking, and hence kitchens, obsolete. Ovens and stoves have been replaced by food replicators that prepare perfectly anything an individual can name. As hostess in the lounge, Guinan ferries food and beverages from the replicator to the customer. With every step, she reinforces the notion that the domestic sphere is the natural space for black women. Her placement in the only public service space is more maddening because viewers know she is more than 600 years old; she alone has experience with situations and cultures that the *Enterprise* encounters. Clearly, because of her experience with the Borg, she deserves a more prestigious—and assisting—position as advisor. Instead, she remains in the kitchen.

Although food replicators have made the need to prepare food and drink "the old-fashioned way" unnecessary, Guinan does do some manual mixing of experimental and exotic drinks. While potentially overlooked or simply interpreted as something for her to do, these actions evoke images of the mammy mixing potions and cures for ailments. Often, Guinan is seen mixing multicolored drinks in oddly shaped decanters. These scenes are filmed in ways that suggest mystical associations. For example, the bar at which Guinan works is lit from below which casts an eerie glow on both Guinan and her potions. Additionally, her concoctions themselves have magical qualities. In "Time's Arrow," she mixes a drink that evaporates on the tongue of the customer for whom it is personally created. In this way, Mammy Guinan is highlighted as having a natural inclination and affinity for preparation, individual attention, and service through stereotypically mythic/mystical abilities.

Along with visually representing and occupying the traditional space of mammy, Guinan exhibits many of the personality characteristics linked with mammy—specifically, she puts her white family first, she is other-centered, and she knows and accepts her place. The promotion of these associations is particularly poignant since characterizations of mammy represent "the normative yardstick used to evaluate all Black women's behavior" (Collins, 2000, p. 73).

Mammy was known and revered for always putting her white family first. Traditionally, mammy loved and cared for her white family before her own, which perpetuated the "ideal Black female relationship to elite White male power" (Collins, 2000, p. 72). At the same time, she posed no threat to either the family or the power structure that supported slavery (Cook, 1995). Although viewers know Guinan has had multiple husbands and children, she has left them to fend for themselves during her employment on the *Enterprise*. Furthermore, like mammy, Guinan's own children exhibit behavioral problems. For example, in "Evolution," viewers learn that her son would not listen to his mother's advice for "several hundred years." However, like a good mammy, her satisfaction comes from "serving in a white family's kitchen" (Turner, 1994, p. 25).

That Mammy Guinan is particularly protective of her white family is evidenced by episodes revolving around the character Q and his mischievousness and meddlesomeness. In "Q Who?" Guinan assumes a cat-like defensive posture while Q is in Ten-Forward. She never lowers her guard as she stands between the crew/her family and the enemy. In the episode "Deja Q," her protective nature is reinforced by the framing of the scene when Q is attacked by an extraterrestrial energy field. As he lies on the ground, viewers see Guinan, dressed in orange robes standing over him as she comments dryly, "Oh how the mighty fall." Her superior position in the frame, as well as the vibrant color of her robes, anchors her solidly as the center of the shot.

Relatedly, Guinan is other-centered. Viewers learn about Guinan through the small bits of herself she reveals as she counsels others. For example, in "Imaginary Friend," Guinan reveals that as a child she had an imaginary Tarkesian Razor Beast as a playmate. She only reveals this, however, to comfort a young girl who also has a make-believe playmate. Her self is known only in the service of others.

Finally, Guinan, like a good mammy, knows her place (Collins, 2000). Although she does not venture out of Ten-Forward often, when she is in public spaces that fall outside of her domain she looks confused and acts displaced. For example, in the episode "Yesterday's *Enterprise*," Guinan goes to the bridge to tell the captain something is terribly wrong. However, when she enters that realm, her steps falter, and she walks reluctantly toward the captain. In addition, she looks around with a confused, and almost pained, look on her face. Clearly, this is not her space and she does not look comfortable being here. Guinan further confirms her contentment with her place by the fact she gives no inclination that she is at all unhappy or desirous of another position. In the episode "The Loss," Deanna Troi loses her empathic powers and doubts her ability to perform her duties as counselor. In a ploy to get Troi to recognize that indeed she is qualified, Guinan acts as if she is interested in the position—a possibility Troi finds incomprehensible. Although Guinan claims the job would come with "better hours" and a "nicer office," viewers know Guinan is not serious by her facial expressions and responses to Troi. During this exchange, it is quite clear that Guinan does not even consider herself appropriate for Troi's position. As expected, Guinan smiles contentedly when, at the conclusion of the episode, Troi announces she is staying.

The most telling aspect of the "mammification" of Guinan is played out by the character herself. Guinan's duties and actions on the ship further promote the reading of her as a traditional mammy. Guinan's duties as mammy are twofold—she is a primary caregiver to children, and particularly Wesley Crusher, and she is an interpersonal advisor to the nonwhite men of the *Enterprise*.

First, and most like stereotypical mammies, Guinan takes care of the white children on the *Enterprise*. There are several instances in which Guinan interacts with the children. For example, when the android Data clones himself and creates a daughter, Guinan allows the girl to work in the lounge ("The Offspring"). When Guinan and three other crew members are turned into children, it is she who helps Ensign Ro through the difficult transition to adulthood ("Rascals"), and when a young girl's imaginary playmate causes trouble on the ship, it is Guinan who comforts the young girl and reassures her ("Imaginary Friend"). Each of these examples illustrates Guinan's role as caretaker/nurturer to the young, white children of the *Enterprise*.

Most illustrative of her role as mammy, however, is her relationship with young Wesley, teenage son of Dr. Beverly Crusher. Wesley is left aboard the *Enterprise* to finish his training during the second season of the show. During this time his mother is given another assignment, so the two spend this crucial year of his development apart. During this same season, Guinan is introduced on the *Enterprise*. The title of the first episode in which Guinan appears, "The Child," clearly defines her in the role of mammy. In this episode, the storylines and interpersonal character development foreshadow Guinan's role as Wesley Crusher's substitute mother/mammy.

In this episode, the young Crusher is upset because he thinks his time to leave the *Enterprise* has come. As he stands in front of the panoramic windows in Ten-Forward, Guinan approaches him and asks him several times whether she can get him something to drink. When he snaps, "No, nothing. That's the third time you've asked me," she replies, "That's what I'm expected to do. Don't you always do what is expected?" After he admits that indeed he tries to do what is expected of him even when it is not what he wants, Guinan wisely advises him, "But sometimes the game is knowing when to consider yourself before others. Give yourself permission to be selfish." She leaves him, an odd look upon his face, contemplating her ironic advice.

Later in the same episode, Wesley goes back to the lounge, and the two begin to build a relationship. During this scene, Wesley is called to the bridge and immediately jumps up. As he is leaving, he turns, smiles at her, and says, "Thanks." Guinan, with her usual air of coyness asks, "For a dish of Duvarian Soufflé?" Wesley shakes his head and replies, "No, not for the soufflé, I meant for the rest." Affirming her place, and her contentment with it, Guinan smiles and tells him, "That's why I'm here." Overall these two scenes, albeit brief, work in combination to establish Guinan as an emotional resource for Wesley. She is, in a very vital and immediate way, taking the place of his absent mother.

Although the mammification of Guinan as exemplified in her enactment of her duties is troubling, more disturbing is the absolute disregard of her input and the undermining of her influence in all areas of her life. One such example occurs at the end of the episode "The Child" discussed above. Arguably, Guinan has played an instrumental role in Wesley's decision to ask Captain Picard to remain on the *Enterprise*. However, once his petition has been made to the captain, Picard takes the request into the realm of white male power. The implications and consequences concerning race are particularly grave.

In the final scene of this episode, all the major characters—Worf, Riker, Picard, Troi, Data and Wesley—are at their stations. As they prepare to "engage," Picard announces Crusher has requested to stay on the *Enterprise* and that "his remaining will cause difficulties for all of us." He goes on to note that since Wesley's mother will not be available, the task of raising the young boy must fall to the crew. At this point, the tasks and responsibilities that are part of raising a child are divided up among the crew along racial lines that are telling as to who may assume responsibility for a white child. Data, as a super-intelligent android, is assigned the task of "seeing to his education," and Worf, as chief of security will see to his safety and will "tuck him in at night." Then, Captain Picard says, "That takes care of the practical, but there's more to growing up than that" and proceeds to assign Commander Riker the responsibility of supervising those non-tangible, nonpractical aspects of growing up. So, as nonwhite crew-members, Data[5] and Worf, are acceptable to perform tasks relating to the young boy; however only the white male can truly see to his upbringing and teach him to be a man. Overall, as later episodes demonstrate, it is Guinan who gives Wesley advice about relationships, school, and broken hearts. However, this final scene reinforces her invisible place within the white family and the subordinate roles nonwhites take in the raising of white children.

As a mammy, however, Guinan must also take care of her black family. Interestingly, on *TNG*, Guinan rarely interacts with the white women on the *Enterprise*. Instead, she is usually seen talking with Data, Geordi, or Worf. Her role as interpersonal advisor to nonwhite men is certainly consistent with the mammy stereotype. She advises Geordi about love ("Galaxy's Child," "Booby Trap") and about working with others ("Hollow Pursuits"). She tells Worf he should find a mate ("Yesterday's *Enterprise*") and advises him what to do with his son ("Redemption"). Even Data gets advice on love ("In Theory") and humor and humanness ("The Outrageous Okona") from her. It seems as though everyone on board can benefit from her wisdom.

Clearly, the character of Guinan enacts the role of mammy on the *Enterprise*. Parkhurst's popular description of mammy written in 1938 could easily be applied to the twenty-fourth-century Guinan. More than a mere mammy, however, as the only female black character with a recurring role on the *Enterprise*, Guinan must function as an amalgamation of traditional black stereotypes and, indeed, be all things to all people. [6] By appealing to the entrenched myths surrounding black women and their responsibilities—particularly their need to sacrifice themselves silently for others—Guinan further limits the agency and sphere of black women generally. The stereotype which most obviously stands in contrast to the mammy, is that of the sexually charged jezebel.

## The Jezebel

A second and equally pervasive myth, which also evolved during slavery, is that of the jezebel. Jezebel is often portrayed as a seducer of white men (Jewell, 1993). Unlike the mammy, she is aggressive with an uncontrollable "sexual appetite" (Collins, 2000, p. 77). Adjectives used to describe the jezebel include: "sexual," "erotic," "seductive," "sensual," "aggressive," and "pushy" (Brown-Givens & Monahan, 2005, p. 93). Although not as evidenced by Guinan, the jezebel has also been characterized as mean-spirited, and rude. Despite the fact she does not enact every quality of the jezebel, Guinan's sexually charged self evokes the stereotype of the jezebel, which in turn, brings with it the baggage of the larger cultural myth. In this, appeals to the jezebel myth undermine any apparent sexual agency of the character and undermine perceptions of the sexual motives of black women.

The jezebel is most often expressed in the private life of Guinan, particularly in her relationship with Picard. A cursory reading of the relationship between Picard and Guinan seems to promote an interpretation of Guinan as empowered. In several episodes Picard seeks her out for advice. For example, in "The Measure of a Man," Picard is given the task of defending Data against the charges that as an android Data is the property of Starfleet. After the first day of the trial, feeling dejected he seeks out Guinan in Ten-Forward and tells her about the situation. Guinan reminds him that "in the history of many worlds there have always been disposable creatures" and provides him with an analogy to slavery that he then uses to win the case. In other instances, she goes to him and tells him, in no uncertain terms, what he must do in a particular situation, which seems to demonstrate her power. For example, in "Redemption II," Guinan goes to the captain's ready room and after asking "Am I intruding?," proceeds to inform him that he needs to address the situation at hand. When he tries to dismiss her by saying, "If you have only a vague intuition," she interrupts and informs him that if she is right, he is "responsible for this whole situation." His next comment indicates he will follow her proposed course of action. She is likewise as curt in "Yesterday's *Enterprise*" when she directly tells the captain what to do when the "old" *Enterprise* is somehow sent forward in time. After hearing Guinan's explanation as to why he must send the *Enterprise* back, Picard responds, "What you're suggesting..." and almost immediately Guinan interrupts and informs him "I'm not suggesting. That ship is not supposed to be here. It has to go back." End of discussion. The influence she is able to exert on the captain is further evidenced by the fact that Picard does admit Guinan possesses an intuition he has "come to trust." Arguably, the fact that she "has the captain's ear" and that he often seeks her out for advice gives the character of Guinan power.

The empowerment of Guinan, however, is undermined by the constant questioning of her credibility by Picard. And, as argued by Turner (1994), the "subtle distortions" are more damaging than the "blatant caricatures" of minority characters (p. xv). For example, in "Q Who?" Guinan urges Picard to leave the area they are in which is inhabited by the Borg. In his personal log, he records, "Despite Guinan's warning, I feel compelled to investigate this sector of the galaxy before heading back." The astute viewer knows this is trouble. Interestingly, the doubt expressed by Picard is not at all entertained by the nonwhite men aboard the *Enterprise*.[7] For example, earlier in "Q Who?" Guinan senses something amiss and calls the bridge to ask whether everything is all right. Despite the fact that her intuition has never been wrong, Riker does not bother to inquire further and just assures her that everything is fine. Still looking troubled, Guinan wanders over to the window and in the process passes La Forge sitting at a table. He notices she is upset and asks her if everything is all right. When she replies, "I don't know" he immediately jumps up and reports that he is "going to check out engineering." Clearly, for La Forge, Guinan's intuition is a valid source of knowledge. Meanwhile back on the bridge, Troi asks, "Where is the Captain?" Only at this point—when the issue is raised by the sexualized, white woman—does Riker begin to entertain the notion that something might be wrong. As he and Troi try to locate the captain, they find he has been kidnapped. By Troi's being the impetus for "sensing" trouble, Guinan's intuitive powers are ignored.

Overall, although the relationship between Guinan and Picard could be read as empowering the black woman since she does advise him as to the best course of action in various episodes, the underlying sexual tension between these characters serves to undermine any power she could have exerted. Although Guinan does not overtly exhibit the sexually predatory behaviors often associated with the jezebel, through innuendo the viewer is encouraged to assume a sexual relationship between Guinan and Captain Picard. Innuendo is apparent in the discourses that take place between the two. In "I Borg," Picard and Guinan are fencing. Critiquing her performance, Picard informs her "You kept letting me inside." Guinan replies, smiling, "Ah, but if I lift up, you lunge right underneath." The double entendre and potential sexual nature of this conversation, when examined in conjunction with other interactions, presents the possibility of a sexual relationship.

The association of Jezebel Guinan with her private spaces is evidenced by her attractive appearance in the movie *Generations*.[8] Guinan is shown in her living quarters; her head is uncovered. Interestingly, this is the only time she is shown without some sort of hat/covering on her head, cuing her as sexual being. Her living quarters, softly lit by dozens of candles, frame and place her in a warm gentle space. Additionally, very complimentary close-ups are used to soften her appearance. The opportunity for the viewer to observe Guinan in her personal living quarters comes as a result of Picard going to see her. She does not look at all surprised at the fact, and indeed looks pleased that the Captain has come to visit her sanctuary.

The relationship between Guinan and Picard is sexually charged, and through the use of coded conversations and strategic cuts, viewers are encouraged to contemplate the extent of their relationship. Both Picard and Guinan have alluded to the fact that their relationship "goes beyond friendship." How far beyond, and in what directions, however, is left unsaid. Viewers are offered little hints that can be read as innocuous or as evidence for their relationship. For example, in "Booby Trap," Geordi asks Guinan what she looks for in a man, and she replies that she "is attracted to bald men." When Geordi asks her, "Seriously,?" she explains that "a bald man was kind to me once. When I was hurting, he took care of me." Although the bald man in question may not have been Picard, he is certainly accurately described as bald. The viewer is left to decide if Guinan's bald caretaker was indeed Picard.

There are several instances in which Guinan and Picard engage in behaviors that hint that a more intimate relationship is happening off-screen. "Time's Arrow II" is one episode that draws on the use of strategic cuts and coy smiles to imply a more intimate relationship. During this episode, Picard and his away team go back in time (to San Francisco in the 1890s) to save Data's life. Surprisingly, "Madam Guinan" is there and unaware of her future aboard the *Enterprise*. The first time she sees Picard, she stops speaking as the music comes up in the background. Picard removes his hat and smiles knowingly at her. With a small smile, she asks, "Do you know me?" He responds, "Very well," and his smile broadens. Later, after she is injured during an explosion, Picard stays with her rather than returning to the ship. As she regains consciousness in the dimly lit cavern, the viewer sees Picard holding her as she lies across his lap. When she questions his decision to stay with her rather than return to his ship, Picard replies, "I didn't want anything to happen to you. You are far too important to me." Madam Guinan, who has yet to meet Picard in the twenty-fourth century, asks him whether they become friends. He smiles at her and replies, "Oh, it goes far beyond friendship." Madam Guinan mumbles that she will have to wait five hundred years and snuggles closer to him. As the camera cuts away, the two are embracing tightly, their faces, mere inches apart. Later in the same episode, once Picard has returned to his proper time, he goes to Ten-Forward. On seeing Guinan, he puts his hands on the bar and simply stands and smiles at her. She smiles back at him, and the two engage in prolonged and serious eye contact. What they know, and what has happened between them, remains a mystery to the viewer; however, enough is implied for even the most casual viewer to assume something is going on.

A more overt and directly sexual interaction between the pair occurs in “Clues.” The episode opens with the crew of the *Enterprise* enjoying some time off. Picard’s voiceover informs the viewer that he will be using his time off to “fulfill a promise to a colleague.” Next, the viewer hears a saxophone begin a bluesy tune as Guinan, dressed for the 1950s in a fitted suit complete with fur wrap, sunglasses, and cigarette holder, enters a run-down office and asks for (private detective) Dixon Hill. Picard has arranged for her to be part of his holodeck fantasy. When the secretary tells Guinan that “Mr. Hill” does not have any appointments, Guinan explains that she is running late because she had trouble getting into the dress. She explains, “It took me a little while to figure out exactly what I was supposed to do with these,” as she pulls up her skirt and snaps her garter. This overtly sexual and playful scene is significant because Picard and Guinan are in a very private space enacting his fantasy. Since he would have programmed the holodeck, he would have designed her outfit, complete with garters. Furthermore, this is the only time Guinan is seen out of her figure-hiding robes, and it is Picard who has chosen her off-duty outfit. Although they are engaging in play together, the viewer is reminded that this is Picard’s fantasy when he is called away, and she declines his invitation to stay and “play out” the program alone.

Although a relationship between Picard and Guinan is alluded to, all of their interactions that could be categorized as intimate happen in private spaces. Some of the most intriguing interactions happen in their quarters. For example, in “I Borg,” Guinan goes to the captain’s private quarters. He answers the door in his V-neck silk robe, drink in hand and asks her to come in. As she enters his quarters, Guinan responds slowly in a low voice, “Are you sure?” She looks him up and down in his robe and remarks, “You look like you’re preparing for a quiet evening.” Picard quickly responds by assuring her, “No, I’m delighted to see you—Can I get you a drink?” as he motions to the couch. She declines the invitation for a drink, and their talk soon turns to practical matters concerning the Borg. At the end of the scene the two are standing toe to toe looking into each other’s eyes as she implores him to talk to Hugh the Borg before using him to “destroy his race.” Although they do not physically touch, the level of comfort each exhibits with the other, the duration of sustained eye contact, and the intimate setting all serve as evidence of a relationship.

A second space in which they interact on a more intimate level is in Ten-Forward; often when they meet in Ten-Forward, they are the only two there. For example, in "Best of Both Worlds," as Picard readies the *Enterprise* and himself for battle, he proceeds to Ten-Forward. As he stands in front of the windows, Guinan, who had been sitting in a corner of the empty bar, asks, "Trouble sleeping?" As he explains that it is a tradition for the captain to tour a ship before battle, she rises and walks toward him. The two walk, side by side, to the bar where they sit and talk about the likelihood of winning the battle. As Picard talks of historical battles, Guinan sits with her hand on her chin and smiles at him. After Guinan responds with an encouraging speech about the survival of the human spirit, the ship is rocked by the beginning of the Borg attack. Immediately, Captain Picard is called to the bridge. After responding that he is on his way, his gaze returns to Guinan. When she returns the stare, he manages a small smile and then heads back to his post. Like many of their interactions, in this instance their communication transcends the need for words.

Although Guinan and Picard affirm their relationship verbally and nonverbally in private spaces, their relationship is not acknowledged in public space. In the episode "The Best of Both Worlds," Captain Picard has been taken captive by the Borg, leaving Commander Riker in command of the *Enterprise*. Guinan goes to him, and during their conversation, she asks, "Did he ever tell you why we're so close?" Riker quickly responds by saying, "No." That his first officer is not aware of potentially one of his most influential relationships speaks to Picard's practice of not revealing his relationship with Guinan in public. The covertness of their relationship highlights the distinction between the public mammy/private jezebel Guinan maintains.

There are several consequences of the privatization of Jezebel Guinan. Most directly, taboos surrounding interracial and interclass relationships are confirmed and upheld in two ways. First, although a sexual coupling is the most plausible reading of Guinan and Picard's interactions, writers do not force the viewer to confront the implications of such a relationship since it is merely implied. Second, because Picard and Guinan imply intimacy only in private spaces, the notion that interracial relationships and those that cross class should be kept secret is validated. They spend significant time together fencing ("I Borg"), playing chess ("Q Who?"), on the holodeck ("Clues"), and talking ("Yesterday's *Enterprise*"); however, all of these things take place while they are alone.

The relationship between Picard and Guinan further serves to validate sexual fantasies around powerful white men and subordinate black women. Jewell (1993) argues that in constructing a system to maintain power, mythic systems are enacted that "explain the differential access and acquisition of various social groups to resources and institutions" (p. 4). In these myths, cultural stereotypes are enacted that prescribe acceptable access and action. In terms of position of power on the *Enterprise*, there can be no two other positions that are as far apart as Captain and host of Ten-Forward. Because of this disparity a relationship between these two characters is made more exotic. The parallel to the master of the house[9] and the servant girl is too obvious to be overlooked.

Overall, the interactions between Guinan and Picard serve to construct her as his private jezebel. That their relationship happens entirely in private, and mainly off camera, allows Guinan to maintain her role as mammy. In this way, the character Guinan reinforces the traditional stereotypes of both the mammy and the jezebel simultaneously.

## Conclusion

Representations of race and gender on *Star Trek: The Next Generation* reinforce many of the negative stereotypes associated with people of color and with women. The character Guinan is a particularly rich source for analysis as she is the point where these stereotypes collide. Her placement as mammy, the subtle undermining of her credibility, as well as the innuendo surrounding her relationship with the metaphoric master of the house work to offset many of the positive characteristics that could be associated with Guinan. Furthermore, the continuation of such stereotypes denies the real lives and experiences of black women.

The opening credits of *TNG* claim to "go where no one has gone before." The changing of the original wording, "go where no man has gone before" indicates an uplifted awareness of gender differences that is, unfortunately, not manifested in the characters and themes of this band of space explorers. Although there are positive qualities associated with Guinan, and despite the fact she can be "read" in an empowering way, the underlying myths to which her character appeals—those of the mammy and the jezebel—work to subvert and limit the role of black women in white spaces.

## Notes

1 An entire issue of *Extrapolation* was dedicated to exploring the use of Shakespeare in *Star Trek: The Next Generation*. Likewise, *Enterprise Zones: Critical Positions on Star Trek* (1996) includes essays that explore "centering subjectivities," "manufacturing hegemonies," and "producting pleasure."

2 Research has demonstrated the significant influence mediated portrayals of African American women have on interpersonal interactions and attitudes (Brown-Givens & Monahan 2005; Monahan, Shtrulis & Brown-Givens, 2005).

3 Interestingly, just as the camera cut away before Kirk and Uhura would have kissed, so too does the camera pan away from the impending interracial kiss of Worf and Troi in "All Good Things" (Bernardi 1998, 130).

4 In a separate essay, the author examines racial projections—and particularly the articulation of racial insecurities through little gray extraterrestrials—as expressed in the narratives of individuals who claim to have been abducted by aliens (Kelley-Romano, 2006).

5 It has been argued that Data's extreme lack of color serves as a signifier of "nonwhiteness."

6 Although Roberts (1999a) claims Guinan could fit the role of the "tragic mulatta." I believe her positioning as jezebel, when juxtaposed with her role as mammy, more completely accounts for her character.

7 For example, in the episode "Redemption," Guinan and Worf talk about what he should do about his son. She advises that he needs to go take care of him immediately. He does not argue, and in the very next scene, the viewer watches Worf ask Captain Picard for time off.

8 That Guinan's personal living quarters never appear in the series is supportive of Collins' (2000) argument that "contemporary mammies should be completely committed to their jobs" (p. 74). Likewise, Guinan's presence in Ten-Forward even when unexpected ("Evolution," "The Measure of a Man," "Best of Both Worlds") speaks to her dedication.

9 Boyd (1996) notes that as captain, Picard privileges "Western, patriarchal culture as a universal standard" (p. 101).

## References

Bernardi, D. L. (1998). Star Trek *and History: Race-ing toward a white future*. New Brunswick, NJ: Rutgers University Press.

Bogle, D. (1997). *Toms, coons, mulattoes, mammies, and bucks. An interpretive history of blacks in American films*, 3rd edition. New York: Continuum.

Boyd, K. G. (1996). Cyborgs in Utopia: The problem of radical difference in *Star Trek: The Next Generation*. In T. Harrison, E.R. Helford, K.A. Ono, & S. Projansky (Eds.), *Enterprising zones: Critical positions on* Star Trek (pp. 95–114). Boulder, CO: Westview Press.

Brown-Givens, S. M. & Monahan, J. L. (2005). Priming mammies, jezebels, and other controlling images: An examination of the influence of mediated stereotypes on perceptions of an African-American woman. *Mediated Psychology, 7*, 87–106.

Casper, M. J. & Moore, L. J. (1995). Inscribing bodies, inscribing the future: Gender, sex, and reproduction in outer space. *Sociological Perspectives, 38* (2), 311–333.

Collins, P. H. (2000). *Black feminist thought: Knowledge, consciousness, and the politics of empowerment*, 2nd edition. New York: Routledge.

Cook, B. R. (1995). See me now. *Camera Obscura, 36* (September), 67–84.

Cox, K. (2005). Reinvigorating Jezebel: Prevailing images of black women on BET. *Edwardsville Journal of Sociology,* 5 (July). Retrieved November 21, 2006 from http://www.siue.edu/SOCIOLOGY/journal/cox.htm

Harrison, T., Helford, E. R., Ono, K.A., & Sarah, P. (Eds.), *Enterprising zones: Critical positions on* Star Trek. Boulder, CO: Westview Press.

Hastie, A. (1996). A fabricated space: Assimilating the individual on *Star Trek: The Next Generation*. In T. Harrison, E.R. Helford, K.A. Ono, & Projansky, S. (Eds.) *Enterprising zones: Critical positions on* Star Trek (pp. 115–136). Boulder, CO: Westview Press.

Jewell, S. K. (1993*). From mammy to Miss America and beyond: Cultural images and the shaping of US policy*. New York: Routledge.

Kelley-Romano, S. (2006). Mythmaking in alien abduction narratives. *Communication Quarterly, 54*, 383–406.

Litwack, L. F. (1982). The blues keep falling. In Janette Faulkner (Ed.), *Ethnic notions: Black images in the white mind* (pp. 15–22). Berkeley, CA: Berkeley Art Center Association.

Monahan, J. L., Shtrulis, I., & Brown-Givens, S. M. (2005). Priming welfare queens and other stereotypes: The transference of media images into interpersonal contexts. *Communication Research Reports, 22* (3), 199–205.

Projansky, S. (1996). When the body speaks: Deanna Troi's tenuous authority and the rationalization of federation superiority in *Star Trek: The Next Generation* rape narratives. In T. Harrison, E.R. Helford, K.A. Ono, & S. Projansky (Eds.), *Enterprising zones: Critical positions on* Star Trek (pp. 10–32). Boulder, CO: Westview Press.

Roberts, R. (1999a). *Sexual generations: "Star Trek: The Next Generation" and gender*. Champaign, IL: University of Illinois Press

Roberts, R. (1999b). Rape, romance and consent in *Star Trek: The Next Generation. Extrapolation, 40*, 21-35.

Sliverblatt, A. (2007*). Genre studies in mass media: A handbook*. New York: M. E. Sharpe.

Thurber, C. (1992). The development of the mammy image and mythology. In V. Bernhard, B. Brandon, E. Fox-Genovese, & T. Perdue (Eds.), *Southern women: Histories and identities* (pp. 87-108). Columbia: University of Missouri Press.

Turner, P. A. (1994). *Ceramic uncles and celluloid mammies: Black images and their influence on culture*. New York: Anchor.

VandeBerg, L. R. (1996). Liminality: Worf as metonymic signifier of racial, cultural, and national differences. In T. Harrison, E.R. Helford, K.A. Ono, & Projansky, S.

(Eds.), *Enterprising zones: Critical positions on* Star Trek (pp. 33–50). Boulder, CO: Westview Press.

Wilcox, C. (1992). To boldly return where others have gone before: Cultural change and the old and new *Star Treks*. *Extrapolation, 33* (1), 88–100.

Wilcox, R. V. (1996). Dating data: Miscegenation in *Star Trek: The Next Generation*. In T. Harrison, E.R. Helford, K.A. Ono & S. Projansky (Eds.), *Enterprising zones: Critical positions on* Star Trek (pp. 69–94). Boulder, CO: Westview Press.

•CHAPTER TEN•

# Of Structures, Stories, and *Spaceballs:* Parody as Criticism of Genre Film and Myth

*Beth E. Bonnstetter**

In their highly influential essay on genre and rhetoric Campbell and Jamieson (1978/2000) declared, "the rhetorical forms that establish genres are stylistic and substantive responses to perceived situational demands" (p. 416). Genre in film is no exception. Although they undoubtedly respond to many different situational demands, one of the more salient functions of genre films' familiar conventions, such as those within science fiction films, is to respond (at least in part) to the perceived cultural need to reinforce values. For example, science fiction comments on contemporary situations in futuristic settings. The *Star Trek* series, both on television and in film, has always concerned itself with the issues of the time. *Star Trek IV: The Voyage Home* (1986) addresses environmentalism and the consequences of hunting animal species to extinction. The values of teamwork, selflessness, and concern for the environment are among the values reinforced. Indeed, genre film can be understood as "a system of conventions structured *according to* cultural values" (Grant, 2003, p. xvi; emphasis added). Viewing the social function of genre in this way is "an idea not dissimilar" to understanding genre conventions as bearers of cultural myth (Grant, 2003, p. xvi).

* Beth E. Bonnstetter (M.A., Colorado State University) is a Ph.D. student at the University of Minnesota, Twin Cities.

Myths are "stories, drawn from history, that have acquired through usage over many generations a symbolizing function central to the culture of the society that produces them" (Slotkin, 1986, p. 70). However, myths are also the narratives that naturalize and perpetuate the ideology of a *particular* worldview (Slotkin, 1986, p. 70), usually the dominant one, and disconfirm other (nondominant) worldviews, thus making the confirmed worldviews part of hegemonic ideology. As hegemonic ideology can be oppressive and/or regressive, viewers' awareness of the perpetuation of myth in film is crucial. Therefore, building on arguments by Gray (2005, 2006) and others, in this chapter I argue that because of its ability to highlight generic conventions, parody is in a unique position to unmask cultural myths in genre film. In unmasking myth, parody criticizes not by condemnation but by recognition. Parody invites audiences to recognize myth and genre and be aware of their cultural roles and cultural conventions, but it does not necessarily ask audiences to adapt a new dominant(ing) ideology. Thus, parody can be a powerful tool for the critic of rhetoric seeking to educate viewers and make them aware of ideology and myth in film. To demonstrate, I offer as a case study Mel Brooks' 1987 science fiction parody *Spaceballs: The Movie* and will show how this film unmasks three major myths that appear in many science fiction films.

## Genre Film and Myth

Altman (1984/2003) explains that films falling under the classification of "genre films" generally have two distinct parts to them. These are "semantics," which are "common traits, attitudes, characters, shots, locations, sets, and the like" (p. 31), and "syntax," or "specific meaning-bearing structures" (pp. 31, 33).[1] In science fiction, the semantic elements may include spaceships, aliens, monsters, and science, whereas the syntactic elements might be humans versus aliens (or other types of wars), the role and dangers of science and technology in human life, and traveling through space/time to meet new life forms. Thus the semantic and the syntactic are dependent on each other to constitute the genre film—without the syntactic, the semantic elements would just be random objects put into a film, and without the semantic, the themes are difficult to articulate. Schatz (1977/2003) explicitly ties genre films and their structure (the semantic and syntactic) to myth. "Considering the genre film as a popular folktale," says Schatz (1977/2003), "assigns to it a mythic function that generates its unique structure" (p. 98). The semantic and syntactic conventions of genre film that are repeated from film to film operate in much the same way as do repeating folktales. The genre film makes cultural narratives (myths) interesting and accessible to the audience through the use of popular syntactic and semantic elements and thus reifies and reiterates those narratives. Thus, genre film "does not *have* a myth; it *is* a myth" (Schatz, 1977/2003, p. 96).

What becomes problematic about the myth in these modern-day folktales is the perception that the ideology within the myth is somehow "universal." Carl Jung and Joseph Campbell have both asserted this universality of myths, Jung through psychoanalysis and Campbell through analysis of the myths of different cultures.[2] This belief in the universality of myth has even made its way into pedagogic tools of film criticism. In his textbook *Anatomy of Film,* Dick (2005), citing the work of Carl Jung, asserts that "making a mythic association [when watching a film] involves remembering a pattern of experience that is universal" (p. 189). However, as Barthes (1957/1972) and Slotkin (1986) point out, myths are *not* universal concepts, but cultural and historical constructs. "The very principle of myth" says Barthes (1957/1972), is that "it transforms history into nature" (p. 129) and "has the task of giving an historical intention a natural justification, and making contingency appear eternal and normal" or universal (p. 142). "Mythological statements," says Slotkin (1986), "always represent the ideology of the moment as if it were the embodiment of divine or natural law, or the reflection of a tradition so ancient that its origins are beyond historical ken" (p. 72). Mythic discourse removes the specific historical context and moment to which the ideology embedded in the discourse responds and makes that ideology seem natural and appear to be "the truth."[3] "Myth does not argue its ideology," Slotkin (1986) warns, "it exemplifies it. It projects models of good or heroic behavior that reinforce the values of ideology and affirm as good the distribution of authority and power that ideology rationalizes" (p. 84). Thus myth not only naturalizes one culture's worldview; that worldview belongs to the dominant power within that culture. Whether myth takes the form of cultural narratives that have been handed down over generations or mythic speech that emerges from a specific cultural situation (see Barthes, 1957/1972), myth serves to reinforce a particular culture's dominant values. The ideologically charged myths that reify the dominant cultural power find in the genre film a subtle means to perpetuate, just as cultural myths did in folktales. As Slotkin (1986) warns, "the genres of modern literary [and cinematic] culture may be seen as bourgeois society's chosen means of transforming and representing its mythic heritage" (p. 79). Gray (2006) adds, "If we take some of the more salient definitions of genre, and remove the word 'genre' from them, they even begin to sound like definitions of ideology...Ideology...—and particularly hegemonic ideology—easily houses itself within genre" (p. 53). Myth, genre, and hegemonic ideology are intimately bound up in one another.

## Parody as an Alternative Criticism of Myth

Although much past film genre research has had to do with exposing the myths of United States hegemony appearing in genre films, genre film study began to move away from such analyses in the 1980s (Grant, 2003, p. xvii). Grant (2003) explains that "previously marginalized voices have been finding spaces from which to speak within the discourse on genre, as in the case of queer readings of popular films" (p. xvii). Grant (2003) defends this move, saying, "While it is true that genre movies tell familiar stories with familiar characters in familiar situations, it by no means follows that they do so in ways that are completely familiar" (pp. xvii–iii). This work is important; it explores "reading as text production...to experiment with a-positional reading...to perform discourse in addition to interpreting it" (Ott, 2004, pp. 206–207), and I do not suggest abandoning it. However, though generic texts may be read in ways that counter their coding, genre films nonetheless are coded for a "preferred" reading (Hall, 1980/2001, p.173), and such a reading can reinforce the myth within those films. Viewers may or may not decide to read these texts in ways other than via the dominant codes (Hall, 1980/2001, pp. 166–176; Perlich, this volume), and therefore discarding the study of the relationship between myth and genre film may be a mistake. Myth does not disappear for all audiences, even if critics choose to find meaning beyond it. For example, many fans of *Star Wars* (1977, 1980, 1983) have developed a "Jedi religion" based on the mythology within, and preferred readings of, the films (Porter, 2006).

That does not mean, however, that traditional analyses of myth in genre film should not change. Audiences may very likely resist ideological myth analysis, because "media texts are an undeniable source of pleasure" and ideological scholarship associates alternative or oppositional reading as either "work" (i.e., laborious) or destructive of the pleasure the text gives (Ott, 2004, pp. 195–197; see also Schiappa, 2006). Buscombe (1970/2003) adds that "the conventions of the genre are known and recognized by the audience, and such recognition is in itself a pleasure" (pp. 22–23). Simply rejecting genre film, or pointing out all the myths and rejecting those, may take away an audience's pleasure, which can lead to resistance to critical viewing of media.

An alternative approach, then, to analyzing myth in genre texts would need to replace the work with pleasure, to make the revealing of the myth in genre fun. Gray (2005, 2006) proposes that parody may be such a tool. Drawing on the work of Mikhail Bakhtin, he proposes that parody "be exploited and deployed to promote media literacy education, for parody can instruct...at the very site of media consumption itself" (Gray, 2005, p. 224). Bakhtin (1981) himself explains that this is possible because, "It is the nature of every parody to transpose the values of the parodied style, to highlight certain elements while leaving others in the shade: parody is always biased in some direction, and this bias is dictated by the distinctive features of the parodying language, its accentual system, its structure" (p. 75). The language of parody usurps its target, uses its target to reveal that target's own construction. Thus, parody "can serve as a powerful foregrounding element, potentially making us more genre literate and more attuned to look at the ideological content that resides within the form itself" (Gray, 2006, p. 55). Moreover, part of parody's effectiveness is that it "does not patronize us or talk down" as much media scholarship does, and "gives the laugher a degree of confidence that can then be used to turn on and analyze a parodic target in more detail" (Gray, 2005, p. 235). "Parody," Gray (2006) continues, "is a teacher, but its method is Socratic, encouraging the audience to make the final and decisive link between criticism and target, rather than merely proselytizing on genre" (p. 47). Indeed, Bakhtin (1968/1984) argues that the laughter in parody "expresses the point of view of the whole world; he [*sic*] who is laughing also belongs to it" (p. 12). Just as much of the pleasure of genre texts comes from recognition of the conventions within a genre, so too does the pleasure of recognizing usurpation of those elements in a genre's parody. Parody, says Bakhtin (1968/1984) is part of the carnival spirit, which "is also directed at those who laugh. The people do not exclude themselves from the wholeness of the world" (p. 12).[4]

Turner (2003) explains that parody uses a genre's "existing codes to *examine* the subject in a humorous way" (p. 48, emphasis added). More than just mirroring another text, "by *constructing* signifiers which inter-relate with past film texts, parody generates textual incongruity" (Harries, 2000, p. 23, emphasis added). This incongruity interrogates "the validity of normative systems" (pp. 23–24). In the specific case of genre parody, when the parody highlights and exaggerates a generic convention, it "makes fun of the way in which a genre works" and "aims to make genre's formal and ideological work considerably harder" (Gray, 2006, pp. 45, 55). Parody calls attention to the very fact that a textual element is a generic convention—something created by people to tell a story. As Gray (2005) puts it, "parodic texts often invoke the correct generic grammar but turn on it and externalize it" (p. 227). By extension, if myth is part, parcel, and perpetuated through generic conventions, parody also draws attention to myth. The myth is no longer embedded in the genre convention, but exaggerated, inverted, brought into the open, and shown for what it really is—a constructed narrative.[5]

## *Spaceballs* and Science Fiction Myth

To demonstrate how a parody of the science fiction genre works to expose its myth, I use as a case study *Spaceballs* and discuss how its parody reveals myth in science fiction. *Spaceballs* tells the story of the Spaceballs, led by President Skroob (Mel Brooks), Dark Helmet (Rick Moranis) and Colonel Sandurz (George Wyner), and their attempt to kidnap Princess Vespa (Daphne Zuniga) of the neighboring Planet Druidia to extort the combination of Druidia's air shield from her father (Dick Van Patten) so that they can steal that planet's air supply. However, their efforts are interrupted by hero-for-hire Lone Starr (Bill Pullman) and his sidekick Barf (John Candy), who rescue the princess but subsequently crash land on a desert moon. There they meet the powerful alien Yogurt (Mel Brooks), who teaches Lone Starr the power of "the Schwartz." The Spaceballs manage to complete their kidnapping and extortion, however, and very nearly complete the destruction of Druidia. Lone Starr rescues the princess, and using the power of the Schwartz, defeats Dark Helmet and the other Spaceballs. He returns Vespa to her father, and on learning he himself is a prince, marries her in a happy ending.

I choose this film as a case study for three reasons. First, *Spaceballs* is primarily a parody of the *Star Wars* trilogy, a science fiction text highly steeped in myth (see Gordon, 1978, pp. 315–320; Campbell & Moyers, 1988; Lawrence, 2006). A clear way to demonstrate how parody can reveal myth is to use a film that parodies a particularly mythic set of films. Second, although primarily a parody of *Star Wars*, *Spaceballs* is also intentionally intertextual. The term "intertextuality" was coined by Kristeva (1974/1984) to recognize how people's previous experiences influence their current experience of a text. Intertextuality that is intentionally placed in film is, according to Ott and Walter (2000),"a textual strategy consciously incorporated by media producers that invites audiences to make specific lateral associations between texts" (p. 430). Intentional intertextuality places a moment from a previous text into the new one simply to initiate audience recognition and memory of the original text. Gray (2005) notes that intertextuality, both intentional and as Kristeva (1974/1984) uses the term, is a crucial critical device in parody (Gray, 2005, p. 227), particularly genre parody (p. 226). By nature, genres themselves are intertextual, because, "after all, we must piece genre definitions together ourselves. We only know a genre as we have experienced it and constructed it" (p. 226; see also Gray, 2006). Viewers bring all of their experiences to a text, but they also must have seen or at least have familiarity with original texts to recognize intentional intertextuality.

What makes *Spaceballs* particularly effective parody is that it does indeed work on both levels of intertextuality: it incorporates not only intertextual *generic conventions* of particular science fiction texts but also *specific moments*, or allusions to those moments. Of course, most of these moments in *Spaceballs* are from the *Star Wars* trilogy, but there are also allusions to other science fiction films. For example, in one scene an alien emerges out of a man's chest much as in a similar scene in *Alien* (1979),[6] and the angles of the shots during the transformation of the Spaceballs' ship frame it to resemble the ship *Discovery* from *2001: A Space Odyssey* (1968). Though these moments in and of themselves may be more iconic than mythic, by reminding us of these iconic moments, *Spaceballs* asks us to recall the myths within the films from which these moments come.

Third, although twenty years old, *Spaceballs* continues to be a popular film. It is shown regularly as a midnight feature—as of this writing, it was shown as recently as February 24, 2007 at Landmark's River Oaks theater in Houston, Texas ("Spaceballs," 2007). A "special edition" DVD set of this film was released in 2005 that contained multiple documentaries about the making of the film among its special features. Most salient, an animated series based on the film premiered on cable network G4 in 2007 (Grossman, 2006).

To show how *Spaceballs*' parody works to reveal myth, I select three prominent U.S. cultural myths that appear in science fiction film, explain them, and discuss how *Spaceballs* uncovers and usurps each myth. The first myth I discuss is a more recent one that was exemplified in the science fiction texts of the 1950s and 1960s: "The Imagination of Disaster" as described by Susan Sontag (1966/1976). The other two myths are Jungian/Freudian myths that appear most saliently in the *Star Wars* trilogy but appear in other science fiction films as well: the monomyth and its American version, and the Oedipal myth.[7]

## The Imagination of Disaster Myth

According to Sontag (1966/1976), the majority of science fiction films of the 1950s and 1960s forwarded a myth she labels "The Imagination of Disaster" (pp. 116–131). The plots of these films usually center on either an invasion by some type of extraterrestrials who have come to destroy humanity or the human discovery of an extraterrestrial race on another planet, who subsequently try to destroy humanity. The Imagination of Disaster, then, involves "the prospect of universal mutilation and even annihilation," coming from "invaders" that have a "non-human" form and want either to obliterate humankind or to impose a "regime of emotionlessness, of impersonality, of regimentation" (Sontag, 1966/1976, p. 127). The Imagination of Disaster myth is tied to a recent "popular mythology [*sic*] for the contemporary *negative* imagination" stemming from "the trauma suffered by everyone in the middle of the 20th century when it became clear that, from now on to the end of human history, every person would spend his [*sic*] individual life under the threat...[of] collective incineration and extinction" brought about by the invention of the atomic bomb (Sontag, 1966/1976, p. 130). Hess Wright (1974/2003) clarifies that when these films were made the disaster was thought to come from the Communist Other, and the myth that these films perpetuate is the fear of "invaders' lack of concern with our traditional values and social structure" (pp. 47–48). Although the Other may no longer be Communist, the idea of the Other destroying U.S. values still exists.

Thus, the Imagination of Disaster, rather than a traditional or Jungian/Campbellian cultural narrative, is a more recent myth that has been inscribed into the culture in the way Barthes (1957/1972) describes myth. The myth is a simplistic, moralistic one cautioning against those whose values may be different from those of a culture, or as Hess Wright (2003) puts it, this myth can "promote isolationism" (pp. 48–49). The Imagination of Disaster appears in more recent films as well: *Independence Day* (1994), *The Matrix* (1999, 2003), and *War of the Worlds* (2005) in many ways reify this myth.

Given the prominence of this myth throughout the history of science fiction film and its centrality in many science fiction films' plots, the fact that the central plot device in *Spaceballs* parodies this myth is no surprise. In *Spaceballs*, the Spaceballs seek to take all of the air from Druidia—a planet that resembles Earth. The Spaceballs are thus an alien threat to an Earth-like planet, much as the aliens threaten Earth in science fiction films bearing the Imagination of Disaster myth. However, *Spaceballs* utilizes a parodic device, inversion, that creates "a signifier which ironically suggests an opposite meaning from its employment in the target text" (Harries, 2000, p. 55) to draw attention to the myth. Rather than monsters or "things," the Spaceballs are human beings, like those on Druidia (and Earth), and are a society that is predominantly white. The central villain, Dark Helmet, though resembling *Star Wars*' robotic/monster-like Darth Vader, has a visor that he lifts throughout the film (as opposed to the very end of the film) to reveal that the character is human(ity). The invaders suddenly resemble the dominant group (and the intended audience), whom the Imagination of Disaster presents as threatened and calls to protect. The Druids, though also appearing to be a white society, are frequently referred in the film as "Druish," a play on the word "Jewish." (Princess Vespa, for example, is at least twice called a "Druish Princess.") Thus not only are the invaders, and the villains, the dominant group; they are invading and threatening to annihilate a group that represents one marginalized in U.S. culture. This, then, is the opposite, the inversion of the plot device in many disaster-oriented science fiction films. Moreover, the Spaceball government is run by *President* Skroob (Mel Brooks), who commands an *army* to carry out operations that are destructive to the Other, the Druids, but that are meant to help their own people. The Spaceballs have used up their supply of air, and now the Spaceball army is invading Druidia to get more to help the people on their own planet. The Spaceballs' government no longer resemble the Communist Other but looks more like that of the United States, which "intervenes" in other countries to "protect the interests" of its own people, or in part because of resources available in those countries. In *Spaceballs*, then, the aliens suddenly are not so alien. They are other science fiction films' victimized culture, and their victims are the Other. Because of this, as well as the portrayal of the Spaceball leaders as humorous (see Eco 1984, pp. 7–8),[8] the audience is asked to identify both with the Spaceballs, whose government and culture loosely resemble the dominant culture in the United States, and with the Other, the Druids, who are framed as the "heroes." The constructedness of the myth is exposed—if the ideology in this myth can be inverted, and the audience can identify with that inversion, then the myth is revealed not to be natural at all.

## The Monomyth

The monomyth dates back to the ancient Greeks (Campbell, 1971; Geraghty, 2005, p. 193), but mythic archetypes and the monomyth become salient with the work of Jung (Zehnder & Calvert, 2004, p. 124) and Campbell (1971). Of relevance to the monomyth is Jung's hero archetype, which forms its basis (Zehnder & Calvert, 2004, p. 124). In some versions of this myth, the hero is called by a "herald" who sends him on a journey into adulthood (Gordon, 1978, p. 320). Gordon (1978) explains that an important element of this myth is the "destiny" of the hero, which helps propel the monomyth forward (p. 321). The hero encounters an advisor (pp. 320–21), acquires a religious tool associated with "the power of the *libido*" (p. 321), destroys some sort of enemy, and "discovers his true identity and attains his true powers, which he realizes were within him all the time" (p. 320).[9] Add Zehnder and Calvert (2004), "The hero comes to some epiphany that allows for a return to and restoration of the Ordinary World," and, "The quest is often personal: for example, to find love, to make peace with the past, to learn humility" (p. 124).

The American version of the monomyth involves "a lone hero motif with an inherent urge to *do good* and be redeemed, to bring about the salvation of those less fortunate" (Geraghty, 2005, p. 193). Moreover, "the desire for liberty, above all else, defines the monomyth and the heroes that follow it. Yet protecting liberty and being redeemed do not save the hero from the obscurity of his isolation, he must leave those people he has helped and go on his way to perhaps help others in distress" (p. 193). Citing Telotte (2001), Geraghty (2005) adds that "the individual was central" in the monomyth—an individual who goes on a journey against his (bias intentional) own environment that "leads to a realization and formation of the self" (pp. 193–194). The American monomyth, then, involves a (male) hero on a quest to do good works for the sake of doing them and to gain knowledge, but one who never stays in one place, leaving those he has helped to continue on his quest. This myth emphases the importance of (male) individual action and self-reliance. Geraghty (2005) cites as examples of media texts with this myth the *Star Trek* series, both television and film, including *Star Trek: Insurrection* (1998), for which the monomyth encapsulates the central plotline (pp. 194–195). Of course, the premier science fiction films in which the monomyth appears are the original *Star Wars* trilogy. The myth also plays a role in *The Matrix* trilogy.

Like the parody of the Imagination of Disaster myth, the parody of the monomyth and its American version is also a central plotline in *Spaceballs.* Lone Starr is the roving hero whose journey involves saving Druidia from the Spaceballs and gaining the knowledge of the mystical Schwartz. However, in another inversion of myth, Lone Starr does not take on his quest for personal knowledge or self-actualization—his herald is money. He is hired to go on his journey, and this occurs regularly, as the film identifies him as a hero-for-hire.[10] As Lone Starr continues on his (hired) quest, it takes a turn, not because of destiny, but because of his own foolishness—he put too little gas in his ship and as a result crashes on a desert moon. After he and his party wander aimlessly through the desert, they are found by the "advisor" figure, Yogurt. Yogurt gives Lone Starr the religious entity, the Schwartz, but his motivation to do so seems to be nothing other than that Lone Starr happens to be there and to forward the plot—a point reinforced by the fact that Yogurt is played by *Spaceballs* director/co-writer/producer Mel Brooks. The libidinous characteristic of the Schwartz is exemplified as Lone Starr battles Dark Helmet—both characters hold the phallic-shaped laser-sword over their phallus to activate it. When Lone Starr completes his mission, returning Princess Vespa to Druidia, he does eventually reject the money, not out of self-actualization or because he must move on, but because he no longer needs it and to show his love for the princess (which has developed as a *result* of the journey, rather than being a motivating force of it). When he finds out that he is qualified to marry the princess, he does not move on in his quest but rushes back to her.

None of Lone Starr's actions, then, are motivated by the quest for individualization or self-actualization or the need to do good and help others. Rather, they are motivated by *selfish* reasons—first money, then (sexual) love. Moreover, much of Lone Starr's success and "destiny" is actually due to convenient plot devices. Lone Starr happens to get hired in his next scene following that in which he finds out he needs the money. His client happens to be the beautiful princess's father, who is willing to pay him to save his daughter (and the car, if possible). He happens to run into Yogurt because his ship crashes on the desert moon, because he conveniently has not put enough gas in his ship. The princess happens to be right where he needs her to be in the Spaceball prison (which has few guards until the "fight scene"). He happens to know the location of the self-destruct mechanism in the Spaceballs' ship (how is not explained). Moreover, those of Lone Starr's "epiphanies" that are not plot devices are blatantly told to him, not discovered. He never "realizes" that the Schwartz is with him all the time, nor does he "realize his true identity"—he has to be told, point blank, by Yogurt. First, when Dark Helmet handily tricks Lone Starr into losing his weapon, Yogurt, having tried to imply to Lone Starr that the power was in him ("use the Schwartz"), only to be answered, "I can't; I lost the [weapon]," has to flatly, explicitly tell him, "It's in you!" Later, when the bookie, Pizza the Hutt, to whom Lone Starr owes money, happens to commit suicide, Yogurt, through a *fortune cookie*, tells him he is a prince so that Lone Starr can abandon the monomyth and marry the princess.

Thus *Spaceballs* undercuts the monomyth in two ways. First, it usurps the driving force behind the American monomyth—Lone Starr is on his quest for love or money not for knowledge or self-fulfillment. Love and money are motivators more common to U.S. culture and practice than the mythic narrative of self-actualization. (How many students go to college to get a better job or find a spouse, versus those who go for knowledge?) Second, it displays the unlikelihood of the "destiny" that propels the monomythic narrative by exemplifying the convenient (contrived) plot devices that drive it forward in film. By showing how the monomythic quest can be for reasons beyond "dong good" (even though doing good may be part of those reasons), and by showing how much of monomythic quests occur because of narrative plot devices not destiny, *Spaceballs* reveals the constructedness of the monomyth. Finally, *Spaceballs* shows the cultural constructedness of the prominent hero archetype—Lone Starr's actions show traces of U.S. culture (desire for money, love, etc.) not mythical, universal characteristics.

## The Oedipal Myth

Although the Greek play *Oedipus Rex* (upon which this myth is based) has far more to it, the crux of the psychoanalytic Oedipal myth is the son competing with his father for his mother. According to Zizek (1999), "When we speak about myths in psychoanalysis, we are effectively speaking about one myth, the Oedipus myth—all other Freudian myths...are variations of it" (¶2). Campbell (1971) elaborates, "The Freudians declare that each of us is slaying his father, marrying his mother, all the time—only unconsciously: the roundabout symbolic ways of doing this and the rationalizations of the consequent compulsive activity constitute our individual lives and common civilization" (p. 355). Using the alternative term "Eden complex," Rabkin (1996) broadens the definition of the Oedipal myth, declaring that Oedipal tragedies and romances are a "basic triadic structure that supports the intergenerational conflict surrounding the transfer of power between two characters of one sex and symbolizes their conflict by a struggle for access to a third character of the opposite sex" (p. 25).[11] Again, though the Oedipal myth is described as a symptom of the psyche in psychoanalysis (of which Jung was a part), scholars such as Burke (1954) and Lyotard (1979/1984) remind us that science, too, is composed of narratives.

Gordon (1978) and Berger (1984) explain that *Star Wars* has particular Oedipal overtones. Berger (1984), writing post *Return of the Jedi* (1983), points out, "To achieve [Jedi] knighthood, Luke [Skywalker (Mark Hamill)] must 'confront' his father, Darth Vader, in a classically Oedipal battle. They fight with light sabers (phalluses) as a hideous and vaguely sexually ambivalent emperor looks on, urging Luke to kill his father and take his place as the emperor's minion" (pp. 71–72). Even Gordon (1978), writing before the father/son Luke/Vader storyline was revealed, recognized Star Wars' Oedipal myth, saying, "We can see here again how [*Star Wars* creator George] Lucas attempts to make this essentially Oedipal fable guilt-free...Vader as destructive devil acts out Luke's patricidal desires" (p. 323). The Oedipal myth is also prominent in such science fiction films as *Blade Runner* (1982) and *Back to the Future* (1985).[12]

*Spaceballs* addresses the Oedipal myth,[13] specifically that found in *Star Wars*, by reenacting the climactic battle scene Berger (1984) describes. However, rather than a sexually ambiguous emperor, they are fighting over the "conquest" of the Spaceball ship, which has been transformed into a gigantic female maid that Dark Helmet has and Lone Starr wants (to destroy). The maid, being a gigantic female ship, is an "exaggeration," another key parodic device (Harries, 2000, p. 83), of the female sexuality in the Oedipal myth (and the Spaceballs' awed and excited reactions to MegaMaid is an exaggerated parody of the male fear of and desire to control female sexuality. The name of the operation the ship undertakes is Operation Vacu-Suck. During an earlier scene, MegaMaid removes the air from Druidia to excited chants of, "Suck, suck, suck!" from the male Spaceballs). During the scene, another inversion occurs: Dark Helmet reveals that he is not Lone Starr's father, or even his brother, but his "father's brother's nephew's cousin's former roommate," which makes them "absolutely nothing." Furthermore, as mentioned earlier, the phallic nature of the Schwartz is revealed and magnified—the two enemies' saber-like weapons emerge after they place them over their own phalluses. Dark Helmet even reflexively comments on this, and the need for phallic domination, as he says to Lone Starr, "I see your Schwartz is as big as mine. Now let's see how well you handle it." In its parody of the *Star Wars* sequence, through the exaggeration of the female figure, the inversion of the expected relational conflict, and the sexual nature of the weapons, the myth is revealed. The phalluses, as well as Lone Starr's and Dark Helmet's comparisons of them (they look down at them shortly after they are generated), and the fight over a giant female ship exaggerate the Oedipal conflict. *Spaceballs* effectively reveals this myth and points out the sexual (and sexist) nature of what on the surface seem to be epic, and antiseptic, narratives.

## Parody as Rhetorical and Critical Device

In a 1994 U.S. Supreme Court ruling involving 2 Live Crew's spoof of "Pretty Woman," parody was cited as having the power to "provide social benefit, by shedding light on an earlier work" (quoted in Gehring, 1999, p. 4). Likewise, the 9th Circuit Federal Appeals Court recently upheld as constitutional parodies of the Barbie brand by Danish band Aqua (*Mattel vs. MCA Records,* 2002) and by artist Tom Forsythe (*Mattel vs. Walking Mountain Productions,* 2003). Thus even the U.S. courts have observed the important rhetorical and critical nature of parody. Dane (1980) explains that "parody deals with literary norms" (p. 153) and, as I have argued, the norms of genre film, but as the courts have observed, and as Harries (2000) adds, "the violation of *any* norm has social and political consequence" (p. 32). Gray (2005) further explains, "Parody encourages viewers to reflect upon media messages and structures" (p. 236). *Spaceballs* reveals the generic structures of science fiction, such as these narratives' central alien-disaster plot devices, its heroes saving the day, and the unnatural, culturally specific, and even overtly sexual notions of epic battles in space. "Parody destabilizes the genre," says Gray (2005), "by simultaneously defamiliarizing it and placing it under a microscope for scrutiny...parody can use a genre's force against itself...by making it hard for readers/viewers to take a genre at full face value, or even seriously" (p. 227). Parody reveals how generic devices (saber and space battles, invading aliens) and the myths that are behind them are constructed, even ridiculously so.

However, unlike satire, which "typically focuses on the inherent weaknesses found in the social order" and is "motivated by specific aims that are either persuasive or punitive in nature" (Harries, 2000, p. 31), the "normative disruption" parody causes can "jar the spectator into questioning norms in general—taking a more critical stance toward 'everyday' normative assumptions" (p. 32). Parody is thus as critical a rhetorical device as satire, but it operates through exemplification and disruption rather than condemnation. If myth deforms its message, transforms "history into nature" (Barthes, 1957/1972, pp. 122, 129), parody points out the deformation by deforming the myth itself. By inverting and exaggerating the generic conventions, it inverts and exaggerates the myth, bringing it to the forefront. What it does not do is try to *punish* the genre or call these myths *inherent weaknesses,* as satire does. Parody has an ambivalent relationship with its text, at once laughing at it but doing nothing to change or correct the codes it laughs at, even reinscribing the codes it hijacks, the myths it points out. Bakhtin (1968/1984) asserts, however, that this is part of the carnival spirit, which is "gay, triumphant, and at the same time mocking, deriding. It asserts and denies, it buries and revives" (pp. 11–12). As Barthes (1957/1972) warns, destroying the myth is not enough: "If one wishes to connect a mythical schema to a general history, to explain how it corresponds to the interests of a definite society...it is the reader of myths himself [*sic*] who must reveal their essential function" by responding "to the constituting mechanism of myth, to its own dynamics" (pp. 128, 129). Parody operates much like Barthes' (1957/1972) dynamic mode of myth reading—it makes the reader aware of the myth and, most importantly, points out that it is not natural but constructed, but it does not attempt to destroy it. Satire, with its aim to correct a particular element of the social order, to destroy one myth in favor of another, seeks only to replace one ideology, one myth, with another.

Because it illustrates rather than condemns, and does so specifically through humor, parody is a powerful tool for the critic. First, because parody does not ask its audience to adapt a new dominant ideology or myth, it is, as explained earlier, carnivalesque—it laughs as part of the world, not in judgment of it. Parody does not hold itself above a mythic idea or ideology; it only reveals ideology and myth's construction. Thus the rhetorical critic can use parodic examples to highlight myth and ideology, and to laugh and criticize as part of a culture, rather than ask his or her audience to replace one myth with another or try to stand outside of culture.

Second, although parody does not fragment ideology, perhaps does not even produce bliss (*jouissance*) as Barthes (1973/1975) or Ott (2004, 2007) describe it, parody nonetheless may be a pleasurable means for audiences to become aware of myth and ideology. Whereas traditional media criticism removes the pleasure from the criticism, and satire calls for a new ideology to replace an old one, parody brings the ideology to the forefront and allows the viewer to make the decision. Explains Gray (2005), "Few of us are likely to feel imposed upon [by parody] in the way we might react to more overtly didactic messages" (p. 234). Audiences who resist other forms of mythic or ideological criticism may be more open to parodic criticism because it makes them laugh. The viewer may feel less like his or her pleasure is taken away, and more like he or she is empowered. "Parody's only prerequisite," says Gray (2005), "is a limited degree of textual and genre awareness, and through this and its tone, parodic humor includes viewers by positing them on the knowing inside, rather than alienating them by positing them on the ignorant outside" (pp. 234–235). Thus studying the parody of science fiction may reveal myths and structures in a way that makes humorous meaning and generates reflection on the generic structures and myths within them in ways that straightforward ideological criticism simply cannot.

Indeed, the parody of myth in science fiction has continued since *Spaceballs*' release and into the new millennium. For example, *Galaxy Quest* (1999) takes on the myths present in *Star Trek* and even within *Star Trek* fandom. *Epic Movie* (2007) parodies several highly mythic texts, including *The Chronicles of Narnia: The Lion, the Witch, and the Wardrobe* (2005) and *Superman Returns* (2006). Even *Spaceballs* itself has come full circle. As mentioned, *Spaceballs* returned as an animated series in the autumn of 2007, and Brooks promised that the new *Spaceballs* would address "society more than having fun with Madonna and stuff" (quoted in Vozick-Levinson, 2006). What can possibly be more intertwined with twenty-first-century society and more ripe for parody than contemporary myth?

## Notes

1 "Syntax" in this case does not mean a formal, linguistic syntax. Whittock (1990) explains that film has, rather than formal syntax, meaning that emerges from the arrangement and composition of images and sound. Syntax as Altman (1984/2003) appears to use it, then, means how the filmic elements are arranged to create themes and meaning.

2 See Campbell (1971) for an explanation of both his work and the work of Jung and other psychoanalysts.

3 For a complete discussion of how myth operates in texts, see Barthes' (1957/1972) *Mythologies*, specifically "Myth Today" (pp. 109–159).

4 Ott (2004) also promotes an alternative to ideological criticism: an erotics of reading that "construct[s] Texts from various fragments of our semiotic laden landscape" and provides each individual with his or her own meaning (p. 207). Such a reading would remove the viewer from his or her immediate experience, and create a bodily meaning so highly individualized that it would overcome hegemony (see Ott, 2007). Although criticism through parody may not operate in quite such a radically liberating way, it may be an intermediate step between Ott's (2004, 2007) solution and current ideological criticism.

5 For a complete discussion on how parody unveils genre conventions, see Gray's (2006) *Watching with the Simpsons* (pp. 43–55).

6 *Spaceballs* even uses the same actor who played the role of the attacked human in *Alien:* John Hurt.

7 Of course, a discussion of each and every myth appearing in science fiction texts exceeds the scope of this chapter, and *Spaceballs* itself cannot possibly address every one. Moreover, many of the myths that I discuss are themselves composites or outgrowths of other myths. The purpose of this chapter is not to be all-inclusive but to demonstrate how parody reveals the myth in genre films' textual conventions. I therefore focus on this selection of myths appearing in science fiction texts and discuss how *Spaceballs* reveals them.

8 Eco (1984) explains that characters who are "humorous" are viewed by the audience as inferior or "animal-like" in some way (i.e., the Spaceballs are morally inferior to the audience because they are stealing Druidia's air), but the audience also "in some way sympathizes with" them. The Spaceballs are funnier than they are evil, and the audience spends a good portion of screen time with them as they overcome their obstacles. The role inversion I describe also makes them humorous.

9 These other characters and elements (enemy, advisor, religious entity) in the monomyth are archetypes as well (Zehnder & Calvert, 2004, p. 124). However, for this discussion, my focus will be on the monomyth itself, and therefore the hero archetype.

10 In many ways this particular usurping of the monomyth was also present in *Star Wars* in the character of Han Solo (Harrison Ford). However, because in its parody of *Star Wars Spaceballs* combines the characterizations of Han Solo and Luke Skywalker into one character, Lone Starr, the monomyth moves from a supporting character to the central one.

11 Rabkin (1996) indicated a difference between the Eden complex and the Oedipal complex in that the former is symbolic rather than literal (p. 25). However, I argue in this chapter that the Oedipal myth is also not literal.

12 Scott (n.d.) discusses at length the Oedipal myth at work in *Blade Runner*, and Gordon (1987) cites the myth's work in his discussion of *Back to the Future*.

13 As do other Mel Brooks films, such as *The Producers* (1968) and *Young Frankenstein* (1974).

# References

Altman, R. (2003). A semantic/syntactic approach to film genre. In B. K. Grant (Ed.), *Film genre reader III* (pp. 12–26). Austin: University of Texas Press. (Reprinted from *Cinema Journal,* 23 (3), 6–18, 1984.)

Bakhtin, M. M. (1981). *The dialogic imagination: Four essays.* (C. Emerson & M. Holquist, Trans.). Austin: University of Texas Press.

Bakhtin, M. M. (1984). *Rabelais and his world.* (H. Iswolsky, Trans.). Bloomington: Indiana University Press. (Original work published 1968.)

Barthes, R. (1972). *Mythologies.* (A. Lavers, Trans.). New York: Hill and Wang. (Original work published 1957.)

Barthes, R. (1975). *The pleasure of the text.* (R. Miller, Trans.). New York: Noonday Press. (Original work published 1973.)

Berger, A. A. (1984). *Return of the Jedi:* The rewards of myth. *Society, 21* (4), 71–75.
Brooks, M. (Producer/Writer/Director). (1987). *Spaceballs: The Movie.* [Motion picture]. United States: MGM.
Burke, K. (1954). *Permanence and change: An anatomy of purpose,* 2nd edition. Indianapolis: Bobbs-Merrill.
Buscombe, E. (2003). The idea of genre in the American cinema. In B. K. Grant (Ed.), *Film genre reader III* (pp. 12–26). Austin: University of Texas Press. (Reprinted from *Screen II* (2), 33–45, 1970.)
Campbell, J. (1971). *The hero with a thousand faces,* 3rd edition. Princeton, NJ: Princeton University Press.
Campbell, J. & Moyers, B. (1988). *The power of myth.* New York: Doubleday.
Campbell, K. K. & Jamieson, K. H. (2000). Form and genre in rhetorical criticism: An introduction. In C. Burgchardt (Ed.), *Readings in rhetorical criticism,* 2nd edition. State College, PA: Strata. (Reprinted from *Form and genre: Shaping rhetorical action,* edited by K. K. Campbell & K. H. Jamieson, pp. 9–32, Falls Church, VA: Speech Communication Association, 1978.)
Dane, J. (1980). Parody and satire: A theoretical model. *Genre,* 13 (Summer), 145–159.
Dick, B. (2005). *Anatomy of film,* 5th edition. Boston, MA: Bedford/St. Martin's.
Eco, U. (1984). The frames of "comic" freedom. In T. Sebeok (Ed.), *Carnival!* (pp. 1–9). Berlin: Mouton.
Gehring, W. (1999). *Parody as film genre: "Never give a saga an even break."* Westport, CT: Greenwood.
Geraghty, L. (2005). Creating and comparing myth in twentieth-century science fiction: *Star Trek* and *Star Wars. Literature Film Quarterly, 33* (3), 191–200.
Gordon, A. (1978). *Star Wars:* A myth for our time. *Literature/Film Quarterly, 6* (4), 314–326.
Gordon, A. (1987). *Back to the Future:* Oedipus as time traveler. *Science Fiction Studies, 14* (3), 372–385
Grant, B. K. (2003). Introduction. In B. K. Grant (Ed.), *Film genre reader III* (pp. xv–xx). Austin: University of Texas Press.
Gray, J. (2005). Television teaching: Parody, the *Simpsons,* and media literacy education. *Critical Studies in Media Communication, 22* (3), 223–238.
Gray, J. (2006). *Watching with the Simpsons: Television, parody, and intertextuality.* New York: Routledge.
Grossman, B. (2006, September 20). *Spaceballs* to become a TV series. *Broadcasting & Cable* [Electronic version]. Retrieved September 21, 2006, from http://www.broadcastingcable.com/article/CA6373939.html?title=Article&spacedesc=news.
Hall, S. (2001). Encoding/decoding. In M. G. Durham & D. Kellner (Eds.), *Media and cultural studies: Keyworks* (pp. 166–176). Malden, MA: Blackwell. (Reprinted from *Culture, media, language,* edited by S. Hall, D. Hobson, A. Lowe, & P. Willis, London: Hutchinson, 1980.)
Harries, D. (2000). *Film parody.* London: BFI.
Hess Wright, J. (2003). Genre films and the status quo. In B. K. Grant (Ed.), *Film genre reader III* (pp. 92–102). Austin: University of Texas Press. (Reprinted from *Jump Cut 1* (May–June), 1, 16, 18, 1974.)
Kershner, I. (Director) & Lucas, G. (Writer/Producer). (1980). *The Empire Strikes Back.* [Motion Picture]. United States: 20th Century Fox.
Kristeva, J. (1984). *Revolution in poetic language.* (M. Waller, Trans.). New York: Columbia University Press. (Original work published 1974.)

Lawrence, J. (2006). Joseph Campbell, George Lucas, and the monomyth. In M. Kapell & J. Lawrence (Eds.), *Finding the force of the* Star Wars *franchise: Fans, merchandise, & critics* (pp. 21–33). New York: Peter Lang.

Lucas, G. (Director/Writer). (1977). *Star Wars: A New Hope*. [Motion Picture]. United States: 20th Century Fox.

Lyotard, J. (1984). *The Postmodern condition: A report on knowledge*. (G. Bennington and B. Massumi, Trans.). Minneapolis: University of Minnesota Press. (Original work published 1979.)

Marquand, R. (Director), Kasdan, L., & Lucas, G. (Writers). (1983). *Return of the Jedi*. [Motion Picture]. United States: 20th Century Fox.

*Mattel, Inc. vs. MCA Records*. 296 F.3d 894 (9th Cir. 2002).

*Mattel, Inc. vs. Walking Mountain Productions, Tom Forsythe*. 353 F. 3d. 792 (9th Cir. 2003).

Ott, B. (2004). (Re)locating pleasure in media studies: Towards an erotics of reading. *Communication and Critical/Cultural Studies, 1* (2), 194–212.

Ott, B. (2007). Television as lover, part I: Writing dirty theory. *Cultural Studies <=> Critical Methodologies, 7* (1), 26–47.

Ott, B. & Walter, C. (2000). Intertextuality: Interpretive practice and textual strategy. *Critical Studies in Media Communication, 17* (4), 429–446.

Porter, J. (2006). "I am a Jedi": *Star Wars* fandom, religious belief, and the 2001 census. In M. Kapell & J. Lawrence (Eds.), *Finding the force of the* Star Wars *franchise: Fans, merchandise & critics* (pp. 95–112). New York: Peter Lang.

Rabkin, E. (1996). Eat and grow strong: The supernatural power of Forbidden Fruit. In G. Westfahl, E. Rabkin, & G. Slusser (Eds.), *Food of the gods: Eating and the eaten in fantasy and science fiction* (pp. 21–38). Athens, GA: University of Georgia Press.

Schiappa, E. (2006). Beyond representational correctness: Rethinking academic criticism of popular media. Unpublished manuscript, University of Minnesota, Twin Cities, Minneapolis.

Schatz, T. (2003). The structural influence: New directions in film genre study. In B. K. Grant (Ed.), *Film genre reader III* (pp. 92–102). Austin: University of Texas Press. (Reprinted from *Quarterly Review of Film Studies, 2* (3), 302–312, 1977.)

Scott, S. (n. d.). Is *Blade Runner* a misogynist text? *2019: Off-World*. Retrieved May 6, 2006, from http://scribble.com/uwi/br/br-misog.html.

Slotkin, R. (1986). Myth and the production of history. In S. Bercovitch & M. Jehlen (Eds.), *Ideology and classic American literature* (pp. 70–90). Cambridge: Cambridge University Press.

Sontag, S. (1976). The imagination of disaster. In M. Rose (Ed.), *Science fiction: A collection of critical essays*. Englewood Cliffs, NJ: Prentice-Hall (Reprinted from *Against interpretation* by S. Sontag, pp. 209–225, New York: Farrar, Straus, & Giroux, 1966.)

Telotte, J. P. (2001). *Science fiction film*. Cambridge: Cambridge University Press.

Turner, M. (2003). Cowboys and comedy: The simultaneous deconstruction and reinforcement of generic conventions in the western parody. *Film & History, 33* (2), 48–54.

Vozick-Levinson, S. (2006, Sept. 25). TV goes to Mel. *Entertainment Weekly* [Electronic version]. Retrieved April 27, 2007, from http://www.ew.com/ew/article/0,,1538836,00.html.

Whittock, T. (1990). *Metaphor and film*. Cambridge: Cambridge University Press.

Zehnder, S. & Calvert, S. (2004). Between the hero and the shadow: Developmental differences in adolescents' perceptions and understanding of mythic themes in film. *Journal of Communication Inquiry, 28* (2), 122–137.

Zizek, S. (1999). From the myth to agape. *Psychomedia: Journal of European Psychoanalysis, 8–9* (Winter–Fall). Retrieved May 6, 2006, from http://www.psychomedia.it/jep/number8-9/zizek.htm.

•CHAPTER ELEVEN•

# Epilogue: The Circle Is Now (In)complete...

*John Perlich and David Whitt*

Myth is, after all, the neverending story

— science fiction author Joan D. Vinge

We ended our prologue with the hope that after reading these chapters on modern mythology you would have, according to Joseph Campbell, been caught by its power. However, being caught by myth should not be the end but the beginning of a longer, and hopefully more rewarding journey for the individual into history, culture, and self. While the science fiction and fantasy texts included in this book might not be celebrated in the same way as Homer's *The Odyssey* or Sir Thomas Malory's *Le Morte d'Arthur*, they reflect the essential functions of myth to inspire, teach, and address social problems. In the new millennium, these problems are global and range from debates over science and technology to religion and politics. In this way modern myths draw upon the stories of the past to provide perspective and hope for the future. Indeed, Stock (1993) argues that humanity's knowledge, vision, and power hold unlimited potential, calling this journey our "mythic saga," which, he believes,

> can do more, however, than inspire us with awe; it holds lessons for us just as the ancient mythologies did for earlier cultures. Some of the lessons are clear: ours is a world of progressive change and adaptation; human society has grown from, and is an expression of, the natural world; all of humanity is joining together and can look toward a limitless future. (p. 244)

The beginning of the twenty-first-century has already seen its fair share of modern myths on the large and small screen, with heroes such as Spider-Man, Superman, and Harry Potter and fantasy epics like *The Lord of the Rings* trilogy, *The Chronicles of Narnia*, and *Eragon*. While the cynic might argue the popularity of these stories is based on Hollywood's worldwide marketing of the latest multimillion-dollar computer-generated blockbuster, we would contend the universal appeal of these texts is their mythic power, drawing upon archetypes of the past that resonate within the individual and throughout culture. In this way mythology is timeless and eternal.

We began our prologue with an allusion to *Star Wars* and it is fitting we end the same way. Astute readers will note that the title of our final chapter is a homage to Darth Vader when he confronts his old mentor, Ben Kenobi, and from behind his breath mask utters, "The circle has been completed...You were once my teacher and I learned much from you. But the time of learning has long passed and I am the master now" (Lucas, 1976, p. 165). In fact, Darth Vader is wrong; the circle is not complete, the dénouement had not yet arrived. Like the *Star Wars* saga, most mythologies are rarely "complete." Instead, mythology evolves and grows with the culture that co-constitutes the essence of myth. For this reason, the circle is a profoundly significant symbol in mythology. In their famous conversation, Campbell and Moyers (1988) explained the significance of the circle as a symbol for infinity:

> Campbell: The circle...represents totality. Everything within the circle is one thing, which is encircled, enframed. That would be the spatial aspect. But the temporal aspect of the circle is that you leave, go somewhere, and always come back....
>
> Moyers: No beginning, no end.
>
> Campbell: Round and round and round. (p. 215)

Mirroring the sentiments of Campbell and Moyers, the editors of this volume note that our circle is both complete and incomplete. We end where we began. We left, we went somewhere (visiting with cyborgs; traveling through fantastic lands; reflecting on the sojourn as we neared the conclusion), and we came back. Our circle might be said to resemble Stonehenge more than a wedding band or the corona during an eclipse. The pillars and stones, set in a circle, represent the contributions of our co-authors. Yet between these rocks and boulders are significant spaces of silence. There are, indeed, many more tales to tell, and while the epilogue is (by custom) a place for endings, we find ourselves musing on new beginnings.

Thankfully, mythology offers scholars a deep and plentiful wellspring of narratives with heroes, villains, creatures, and quests as vivid and rich as the cultures they inspired. This is clear when you consider Joseph Campbell's answer to Bill Moyers' question (1988):

> Moyers: What do you make of that—that in very different cultures, separated by time and space, the same imagery emerges?
> Campbell: This speaks to certain powers in the psyche that are common to all. (p. 217)

Although the individual stories are idiosyncratic, the essence of the stories remains the same. This "neverending story" that science fiction author Joan D. Vinge refers to in our epigraph is an essential part of who we are, where we came from, and who we will become. And as the years turn to decades, decades into centuries, and centuries into millennia, what will always remain is the story, our story, the story of the human spirit.

## References

Campbell, J. & Moyers, B. (1988). *The power of myth*. New York: Doubleday.
Lucas, G. (1976). *Star wars: From the adventures of Luke Skywalker*. New York: Del Rey.
Stock, G. (1993). *Metaman: The merging of humans and machines into a global superorganism*. New York: Simon and Schuster.

# Index

## •S•



## •T•



## •U•



## •V•



## •W•

## •X•

Toby Miller
*General Editor*

*Popular Culture and Everyday Life* is the new place for critical books in cultural studies. The series stresses multiple theoretical, political, and methodological approaches to commodity culture and lived experience by borrowing from sociological, anthropological, and textual disciplines. Each volume develops a critical understanding of a key topic in the area through a combination of thorough literature review, original research, and a student-reader orientation. The series consists of three types of books: single-authored monographs, readers of existing classic essays, and new companion volumes of papers on central topics. Fields to be covered include: fashion, sport, shopping, therapy, religion, food and drink, youth, music, cultural policy, popular literature, performance, education, queer theory, race, gender, and class.

For additional information about this series or for the submission of manuscripts, please contact:

Toby Miller
Department of Cinema Studies
New York University
721 Broadway, Room 600
New York, New York 10003

Lightning Source UK Ltd.
Milton Keynes UK
UKHW021357070722
405523UK00004B/86

9 781433 100956